'*Citizens* is a breath of fresh air amidst deep concern about the future of democracy. It empowers and calls us all to action to be the democratic change we want to see in our communities, and in doing so offers a powerful vision for the transformation of our institutions.'
Marietje Schaake, International Policy Director, Stanford University Cyber Policy Center, and Author, Democracy.com: How to Stop Tech Companies from Ruling our Digital Lives

'Jon Alexander's New Citizenship Project speaks to action in the real world but has an immensely solid base in ideas as expressed in words, the meanings and applications of which he first teases out and then rams home with elan and gusto in this bravely inspiring book. The range of reference in both time and place from ancient Greece to the latest democratic community project is spellbinding and salutary, as is his injunction to get out there and make a positive, activist difference.'
Paul Cartledge, Emeritus Professor of Greek Culture, Cambridge University

'Jon Alexander's work has had a significant influence on my understanding of the role employees, investors, and customers could play in shaping the role of business in society. To harness this potential, leaders must first see people as the creative, empathic, collaborative members of communities that they truly are: as *Citizens*. In this engaging book, Jon lays out his full vision for how this mindset shift can transform not just business, but NGOs and governments too.'
Alex Edmans, Professor of Finance, London Business School

'*Citizens* is a powerful provocation for our times, one that draws on humanity's deep capacity for cooperation to offer a new way forward. Highly recommended.'
Nichola Raihani, Professor of Evolution and Behaviour, University College London, and author, The Social Instinct: How Cooperation Shaped the World

'I couldn't agree more with Jon's diagnosis and prescription for how we need to revitalise our civic institutions and communities. The wonderful thing is that he not only gives us hope but more importantly he lights a pathway to make this new paradigm a reality through the years of deep work, thinking and action that have formed the basis of his book.'
Jason Stockwood, Vice Chairman, Simply Business, and Chairman, Grimsby Town Football Club

'Society is like an out of control house party – eating, drinking and consuming everything. Jon is the organiser of the campfire gathering behind the party. It's calm and welcoming and you won't want to leave. In *Citizens*, Jon and Ariane show how to leave the burning house of the Consumer Story and join the campfire that is the Citizen Story.'
Stephen Greene, CEO of RockCorps and founding Chair of National Citizen Service (UK)

CITIZENS

Jon Alexander
with Ariane Conrad

First published by Canbury Press 2022
This edition published 2022 (second reprint)

Canbury Press
Kingston upon Thames, Surrey, United Kingdom
www.canburypress.com

Printed and bound in Great Britain
by CPI Group (UK) Ltd, Croydon
Typeset in Volkorn/Interface
Cover: Daniel Benneworth-Gray

This is a work of non-fiction

FSC® helps take care of forests for future generations.

ISBN
Hardback: 978-1-912454-84-6
Ebook: 978-1-912454-85-3

FOREWORD

Is there any way out of the mess we're in?

You certainly wouldn't think so from reading the papers. The richest people in the world are already scrambling onto the lifeboats, be they called New Zealand or SpaceX. And the poorest are dying in increasing numbers as they flee their war-torn, climate-ravaged countries. Meanwhile official politics seems either impotent or positively malign, a well-oiled machine infallibly finding the worst and pushing them to the top of the heap. Most of the graphs seem to be pointing in the wrong directions: going up when they should be going down, or plummeting when they should be rising.

Yet there is hope. Because something is happening. There is a coalescence. A different story is rising and ripening. It is a story of who we are as humans, what we are capable of, and how we might work together to reimagine and rebuild our world.

This story doesn't show up on the media radar because that radar is resolutely pointed in the wrong direction. It's expecting the future to be produced by governments and billionaires and celebrities, so its gaze is riveted on them. But behind their backs, the new story is coming together. It is slower, more diffuse, and more chaotic, because it is a story of widely distributed power, not of traditional power centres. As Jon Alexander and Ariane Conrad name it in this vitally important book, this story is the Citizen Story. When you see the world through the lens of this story, you see that there is a revolution in progress. The people are organising, not only in grassroots movements but also, crucially, inside the very institutions and organisations that are currently failing us. The people are starting to feel their power, and they're making for the engine rooms.

❀❀❀

Stories matter. They can trap us, but they can also inspire us.

The stories we tell shape how we see ourselves, and how we see the world. When we see the world differently, we begin behaving differently, living into the new story. When Martin Luther King said: 'I have a dream,' he was inviting others to dream it with him, inviting them to step into his story. Once a story becomes shared in that way, current reality gets measured against it and then modified towards it. As soon as we sense the possibility of a more desirable world, we begin behaving differently, as though that world is starting to come into existence, as though, in our minds at least, we're already there. The new story becomes an invisible force which pulls us forward. By this process it starts to come true. Imagining the future makes it more possible.

Sometimes this work of imagination and storytelling is about the future, as in Dr King's story. Art can play this role: what is possible in art becomes thinkable in life. We become our new selves first in simulacrum, through style and fashion and art, our deliberate immersions in virtual worlds. Through them we sense what it would like to be another kind of person with other kinds of values. We rehearse new feelings and sensitivities. We imagine other ways of thinking about our world and its future. We use art to model new worlds so that we can see how we might feel about them.

Sometimes, though, the work of imagination is about reappraising now. That's what this book aims to achieve. It is about shifting the perspective so we see the world we are in now with new eyes.

❀ ❀ ❀

Climate change is the biggest threat that humans have ever faced, and the systems and institutions that govern our world today are proving themselves unable to respond to it. The two dominant narratives today are China – a Subject state with centralised power and deep surveillance – and Siliconia – a Consumer state with centralised power and deep surveillance. Neither offers a credible path to a sustainable, equitable, beautiful world.

That's the bad news.

The good news is that the climate emergency has given rise to the largest and most broadly based social movement in human history. That movement has a billion roots, a billion people who set out on their own or in small groups and are now starting to meet up and look around – surprised and exhilarated – at each other. This huge, global movement is starting to become self-aware, to understand its size and power. It's the critical moment, for social revolutions happen in two phases. The first is when everybody realises the current system isn't working any longer. The second is when everybody realises that everybody else has realised it too. Then the coalescence begins – the movement gels. The vast majority of people realise they're on the same side. The people come together, and there's no stopping them.

This coalescence will take us all somewhere new. The future won't be dominated by either of the two narratives currently under discussion, but by what this book calls The Citizen Story.

The Citizen Story is about the empowerment of us all to co-invent and nurture our own futures. The Citizenship of this book is not about the passport we hold, and it goes far beyond the duty to vote in elections. It's a state of engagement, more verb than noun. We look around, identify the domains where we have some influence, and we roll up our sleeves and make things happen.

People are losing faith in the old neoliberal story – that everything would work out for the best if we just let the market do its work and didn't interfere. We've seen where that has got us. People who care about the world their grandchildren will inhabit are watching wealth concentrate into fewer and fewer hands and those hands aren't building a viable future. Marginalised people

– women, gender non-binary, people of colour, the colonised, the disabled, the poor - are all demanding inclusion and consideration and respect in our global civilisation. All these people are asking to be Citizens. These people don't want the Consumer Story any more. They want the Citizen Story.

What we're realising together is that all the ever-louder cries for freedom and justice are entangled. They're all part of the same revolution. Gender rights, indigenous rights, ethnic rights, animal rights... they're all part of the same fundamental rethink about relationships to each other and to the planet. All life on Earth is an ecosystem, and it could be an ecosystem of generosity, a virtuous circle. If we improve things at any point we improve them at all points. That is what this movement and this moment are really about. To save the planet we have to do so many things; but if we do, we will end up in a much better place than we are now.

In naming the Citizen Story, and its Subject and Consumer counterparts, Alexander and Conrad have made a vital contribution. They have made a new present and a new future real, making it possible to see a bigger idea of what is and what could be. They have also made it possible to discern and distinguish the Citizen path from those of the Subject and the Consumer, enabling us to leave those behind once and for all. Equipped with the ideas in this book, what I now see is not just a climate emergency: it's a civilisational opportunity.

Brian Eno
November 2021

For Jane

CONTENTS

OPENING

This book reveals what is quite possibly the only thing that can save us now: the Citizen inside every one of us. Which is to say: there is still hope. The future is up for grabs.

In order to survive and thrive, we must step into the Citizen Story. We must see ourselves as Citizens – people who actively shape the world around us, who cultivate meaningful connections to their community and institutions, who can imagine a different and better life, who care and take responsibility, and who create opportunities for others to do the same. Crucially, our institutions must also see people as Citizens, and treat us as such. When they do, everything changes. If we can step into the Citizen Story, if we can transform our institutions, we will be able to face our myriad challenges: economic insecurity, ecological emergency, public health threats, political polarisation, and more. We will be able to build a future. We will be able to *have* a future. That is what is at stake here.

The Citizen Story is already present, just beneath the surface of our society. It has been taking shape in the national and regional politics of Taiwan, Ireland, Iceland, Belgium, and beyond, where governments are embracing a role that is more about coordination and facilitation than central command. It's been emerging in Barking and Birmingham; Paris in France and Porto Allegre in Brazil; Calgary in Canada; and little Frome in Somerset, where mayors and local politicians of contrasting ideology, age and skin colour have been tapping into the energy and ideas of their people to resuscitate their cities and towns. It's been building in refugee camps and shanty towns, in big charities and social movements, in museums, in schools, and even in businesses, as all sorts of organisations and institutions find success through involving, rather than telling or selling. People everywhere are joining in, and creating the possibility of a truly sustainable, inclusive, joyful future as they do so.

All this is happening. But so far, it is a long way from enough. As things stand, too many of us are not engaging in the world around us. We have – understandably – lost faith in our institutions, most especially our governments, and so we keep our focus close, strive to get our own needs met in a world that seems to be growing harsher and less safe by the day. We are living inside what I call the Consumer Story, which tells us we are entitled and passive: we are to be sold to and served.

The challenge is not that the Citizen Story is complicated to articulate or difficult to evidence; it is simple, rooted in deep truth, and emerging everywhere. But it is hidden by the dominance of the Consumer Story. The organisations and institutions of our society – not just our businesses but our governments too, and every other sector – reinforce this story over and over again, every single day. They have done so to the extent that it is often mistaken for fundamental human nature. The Consumer Story has come to feel inevitable, unbreakable. But it is not the true story of humanity; it is simply the story of self and society that most of us – almost all of us – have been brought up within.

. I am no exception. Having spent the first decade of my career in London's advertising industry, I'm more like the rule.

LEAVING ADLAND

I started my working life at 151 Marylebone Road in London, the offices of Abbott Mead Vickers BBDO. It was Thursday 25th September 2003 and I was 21. Standing before the doors of AMV – considered one of the world's most prestigious and successful architects of human aspiration and behaviour, much lauded for its power in making brands into household names – I was not only seeking my fortune but ready to make what I then understood as my contribution.

Advertising was a dream job in the world of that time, the only world I'd ever really known. I was two years old when the big consumer superbrands arrived, when in the single symbolic year of 1984, Apple launched the Macintosh, Nike sold the first Air Jordans, and Virgin Atlantic broke open the skies. I was 15 when Tony Blair and New Labour heralded Cool Britannia, told us things could only get better, and were very comfortable with people getting filthy rich. And I was 19 and at university when, two years and two weeks before my first day at AMV, the twin towers of the World Trade Center came crashing down. In the wake of 9/11, George W Bush, the US President, Blair, the British Prime Minister and the rest told us that what we the people could do – what we should do – was buy stuff. Keeping the growth of the economy going was how we could show those who aimed to terrorise us that we were not afraid. We would keep calm and carry on shopping. And I would help.

My first boss at AMV had bought the story in a big way. He gave me a copy of Francis Fukuyama's *The End of History*, the widely influential book that declared that humanity had reached its zenith in liberal capitalist democracy. My boss interpreted the book as meaning that advertising was the cutting edge of the only future going. And he defined my new job as a Graduate Account Manager like this: 'The average consumer sees somewhere between 1,500 and 3,000 commercial messages a day. We have to cut through that. Our job is to make our clients' messages the ones that stick.' At first, I took this task at face value, aimed for excellence within the rules of the game, and often succeeded. Both AMV and the next agency I worked for won Agency of the Year while I was there. One year, one of my projects even won *Brand Republic*'s Big Creative Idea of the Year.

But while I played the game, I felt uncomfortable almost from the beginning. It was partly the day-to-day: for all that I'm still close to many of my former colleagues, there were others who did things in their personal lives that I wanted no part of (think *Mad Men* and more). It was partly my belated but growing political awareness. I read *No Logo*, Naomi Klein's deconstruction of my industry: according to her, the principal impact of my long hours of work was to equate products with lifestyles and with identity, leveraging 'cool' and youthfulness and desirability, and to help corporations obscure the suffering those products caused to workers and the environment.

I watched Adam Curtis's classic documentary series *The Century of the Self*, and learned how, using the psychology behind desires and aspirations, Sigmund Freud's nephew Edward Bernays and the field of public relations professionals he had inspired had not just increased their clients' sales; they had shaped the course of 20th Century politics. Viewing politics as just another product to be sold to the public, they had advised numerous administrations, to the extent not just of shaping electoral campaigns but even launching wars. The year I entered the workplace was the same year that Bush and Blair led the invasion of Iraq, bringing more than a million people out onto the streets of London in protest. I could only look the other way for so long.

Mostly, though, it was the climate. This was the one I couldn't hold alongside my work, the arena where the clash was too obvious and too direct. Doing a good job meant selling more stuff, yet ever-increasing consumption was obviously unsustainable. For a while I accepted the lullaby of 'conscious consumption,' and did my best to sell greener stuff: Eurostar instead of flying, fair-trade chocolate, and so on. This even, briefly, made my choice of career feel good again: where better to work than in advertising, if what we needed to do was modify the nature of consumption? But my relief was fleeting, not least because for every pound spent promoting the ethical or environmental alternative, even just within the walls of the agency I was working at, another five or ten or even 20 were at the same time pushing the opposite.

Around this time, the words of my first boss began to haunt me. A question took shape in my mind:

What are we doing to ourselves when we tell ourselves we're Consumers 3,000 times a day?

What if the impact wasn't just about the stuff we were consuming, and the material impacts of that (as important as those things are), but something much more pervasive? What my advertising work was part of, I began to see, was the telling – and retelling and retelling, thousands of times a day – of a story about who we are and what we're capable of. A story that creates not only material but also psychological and even spiritual problems, and at the same time limits what we believe ourselves to be capable of doing in response to those problems.

The Consumer Story, as I would come to think of it.

This story is morally justified by a vague theory that everyone of us pursuing our own self interest will add up to collective interest. It is a story in which human society is essentially and necessarily a competition, since we lazy, selfish humans can be driven to act only by the competitive imperative, which arises from primal instincts to protect and pass on to our own. The institution in this story functions above all to serve us. We feel entitled, and believe ourselves self-reliant and independent, the creators of our own destiny. Because 'the Consumer is King,' we believe that we are in charge. The glittering array of available choices makes us forget that there might be possibilities beyond the menu, or damage inherent in the very dynamic of our choice between them. That's true not just in stores with crowded aisles, but in politics with its well-packaged candidates, dating platforms that seek to sell us our new partner, charities that compete to offer 'benefits' for our support, and on and on. Intoxicated by the power to choose, we don't realise that the real power lies in creating the menu – even in deciding whether there is a menu at all.

This revelation flipped a painful switch in my head. Suddenly, I saw working in advertising not as promoting values I could and should believe in and be proud of; but as preaching the Consumer Story. This was a story I no longer believed in. In fact, it was a story I was appalled by.

From this point on, working in advertising was torture. Things got worse – arguments, shouting, tears, at work and at home – and then worse still. Late one evening, standing on the platform at Oxford Circus underground station, I stared at the floor as one train came and went and then another, waves of nausea passing over me. I honestly don't remember whether I was contemplating taking my own life, though I may have been; all I remember is the nausea.

Another train passed, and then I started retching, and then threw up. It happened every day for a week. At the end of the week, I resigned. I really, really hated myself.

Ever since those awful days, I have been following that initial question that spurred my unravelling: What are we doing to ourselves when we tell ourselves we're Consumers 3,000 times a day?[1] In doing so, I've come to see 'Consumerism' not just as a term to be used in lamenting excessive material consumption, but as a deep and powerful story that shapes who we think we are, what we think we are capable of, and how we relate to one another. That Consumer Story is entrenched and embedded in each and every one of those 3,000 daily messages, and much else besides.

Once I had stepped out of advertising and was able to see more clearly, I developed another question, one that arose not from staring obsessively down at the trash-filled gutters, but instead looking up and around at the compassion, creativity and brilliance exhibited daily by humanity, in spite of the ubiquity of the Consumer Story. I began to ask:

What would it look like if we put the same energy and inspiration that currently goes into telling ourselves we're Consumers, into building our agency as Citizens?

In 2014, I co-founded a company, the New Citizenship Project, to serve as a vehicle for this inquiry. Together with my colleagues, I developed an understanding that in the early 21st Century, the Consumer Story is collapsing under the weight of its own contradictions, and the Citizen Story is emerging. People are dissatisfied with being mere Consumers, yearn for deeper agency even though we lack the words to express it, and have an innate if imprecise sense that authentic participation holds the key to a brighter future. But the most powerful insight at the heart of NCP's work – and of this book – is that this shift in story is already happening. It is not something that needs to be created or insisted upon or campaigned for; rather, it is something that needs to be named, nurtured and accelerated. It needs to be revealed, not just in ourselves but in organisations of all shapes and sizes, so that all of us can step into it.

As my distance from my time in advertising increased further, I also began to see the Consumer Story in a new light: less as some sort of evil conspiracy, and more as a story that had its moment and served its purpose, and now must be superseded. I began to see how much it had offered and how it had retained such appeal, by seeing it not just in contrast to what was now possible, but in the context from which it had arisen. For there is a third story, one that had dominated before: the Subject Story. Seen in this light, the emergence of the Consumer Story had in fact been liberating, holding its own promise of a new golden age.

In the Subject Story, as in 'subjects of the king,' the greater part of humanity is infant-like and guileless. Our role as the little people is to be ruled over and kept safe by the hopefully benevolent Great Man in charge, the one ordained by right of birth and by god.

Societies and organisations are paternalistic, hierarchical and fixed, with the inherently superior few at the top. They tell the rest of us what to do, declaring our duties. For our part, the right thing to do is obey. The bargain is protection, security, and the maintenance of order, both physical and psychological. When the Subject Story was dominant, it told us that the God-given few knew best, so when we did what they said, we would be protected; life would be as good as could realistically be hoped for.

At the New Citizenship Project, we developed the table on the following page to compare and contrast the key dynamics of each of these three stories.

SUBJECT	CONSUMER	CITIZEN
DEPENDENT	INDEPENDENT	INTERDEPENDENT
TO	FOR	WITH
RELIGIOUS	MATERIAL	SPIRITUAL
DUTY	RIGHTS	PURPOSE
OBEY	DEMAND	PARTICIPATE
RECEIVE	CHOOSE	CREATE
COMMAND	SERVE	FACILITATE
PRINT	ANALOGUE	DIGITAL
HIERARCHY	BUREAUCRACY	NETWORK
SUBJECTIVE	OBJECTIVE	DELIBERATIVE

As I came to see the Consumer Story from this perspective, I was able to recognise its power and appeal. From what came before – blind obedience as Subjects – it felt like a breath of fresh air, liberation and empowerment, placing the individual at the centre of the universe. This also led to the realisation that it was nowhere near as entrenched and inevitable as I had thought. After all, it has been the dominant story for only 80 years or so; its very dominance is evidence that stories can and do change.

And so to now. The Citizen Story is emergent. The Consumer Story is failing, but tenacious, and the Subject Story is resurgent. The struggle for our next story has grave consequences, as recent history has painfully illustrated.

BRITAIN BETWEEN STORIES

There was a specific day when I knew I had to write this book, after several years spent mulling. It was Saturday 10th May 2020, the day Prime Minister Boris Johnson announced the 'roadmap' out of Britain's first coronavirus pandemic lockdown, together with a change in the key message to the nation.

Instead of 'Stay Home. Protect the NHS. Save Lives.' we were now asked to 'Stay Alert. Control the Virus. Save Lives.'[2]

This wasn't a public health moment, I realised; it was a political moment. It was a moment that could only be understood through the lens of the stories of self and society: the stories of Subject, Consumer, and Citizen. It was a change aimed at reframing the crisis and the actors in it not to save lives, but to preserve power.

Until that day, and from the moment Johnson had belatedly acknowledged the severity of the threat, Covid had been cast as an intentional attacker, complete with war metaphors and even comparison to 'an unexpected and invisible mugger.'[3] We the public were the weak and hapless victims who needed to be told what to do; Johnson and his government the strong, all-knowing heroes who would save us. Pure Subject Story.

But by early May that had begun to fall apart. Commentators and academics were gaining an audience as they pointed out that this was not so unexpected; indeed, some kind of pandemic had been widely predicted. Government had not been strong and all-knowing: it had responded too late, and the 'led by science' trope was in the process of being dismantled.

When Britain exceeded Italy's death rate earlier that week, and on almost the same day Germany and other nations began to

21

lift restrictions, the Subject Story could not hold.

And so back came the Consumer. With 'Stay Alert,' each of us now needed to take personal responsibility for dealing with Covid, and getting back to normal as best we could. In the Consumer Story, the right thing for government to do is to step back and get out of the way, because people are best left to look out for themselves. We are individuals, there is no such thing as society. The dark corollary of course was that if you caught the virus after this point, it would be your fault – because you would not have stayed sufficiently alert. Johnson and his government were pushing responsibility for whether we lived or died from themselves onto each of us, but without giving us any of the tools or information we needed to make the right decisions – all of the responsibility, none of the power. The intention? We would blame each other, not the government. An outbreak in Manchester or Glasgow or wherever would have been caused by individuals being less responsible than me.

What was crushed in this moment – or at least suppressed – was the surge in the Citizen Story that had flowed from the moment news of the virus had hit our shores. So many of us in so many ways had thrown ourselves into helping one another: the mutual aid groups, 'ViralKindness' postcards, the 750,000 people who tried to sign up as National Health Service First Responders inside 48 hours.[4] Yet all this was perceived either as a threat to the power of government, or not perceived at all, unintelligible through the Consumer/Subject lenses that filtered reality. The Citizen Story was not sufficiently named or formed or understood to offer an alternative.

Yet the desire was there. I wrote about this moment that very day, a piece that was read over 600,000 times.[5] I knew then I had to write this book, to do the deeper work of interrogating and bringing all this together into a story that can hold us.

Over the course of the summer, Eat Out To Help Out and front page headlines told Britons to 'go shopping.' The rhetoric of personal responsibility took hold, and has continued ever since. Johnson even claimed that 'greed and capitalism' brought Britain its relative success on vaccinations.[6] This has led some to feel that the moment of opportunity for a new story has passed, that the portal has closed. I don't agree with that. The Consumer Story is fundamentally broken. The trigger might be this pandemic or another, the onset of the climate emergency or the next banking collapse, surges in energy from Black Lives Matter or Extinction Rebellion or another movement, or something else entirely. But the old story will crack open again, and there will be more chances to step into the new.

So how do we ensure that the next era belongs to the Citizen, which is to say, to all of us, at our most powerful?

UNLEASHING OUR POWER

The primary focus of this book is not about how to build a movement, or organise a community, or even be a Citizen. That work is well underway, and there exist many excellent resources to support it.

Instead, it is about how we make the transition to a Citizen society. How we recognise the Consumer Story and get it out of the way. How we establish the Citizen as the dominant story of society. The critical task is to redesign the organisations and institutions of our society. We are Citizens by nature. But as things stand today, almost every interaction with almost every organisation – business, government, even charity – conditions us to believe otherwise. While our organisations remain trapped in the Consumer Story, they will keep us trapped too. As such, to free ourselves, we will need to set them free too.

This work is starting to take shape. More and more organisations and institutions, across the world and in every sector, are being transformed or replaced as the Citizen Story emerges and spreads. These organisations can be recognised by the presence of two key characteristics. The first is philosophical. Citizen organisations are rooted in a deep and resilient belief in humanity: a belief that, given the right conditions, it is human nature to want to contribute positively and meaningfully to shape the communities and societies we are part of; and that the capability to make such a contribution is also universal. If you analyse the language of Citizen organisations – their websites, the speeches of their leaders – you can see that they don't so much seek to be trusted or admired by people; rather, they start by trusting and believing in people.

The second follows naturally from the first, and is practical. Because they believe in the contribution people can make, Citizen organisations adopt core processes that are designed to make those contributions possible. They are open and transparent; they show their working; and they invite participation at every available opportunity. As a result, people's massive loss of trust in organisations is flipped: every interaction with every organisation becomes characterised by their trust in us. One indicator of the emergence of the Citizen Story is the sheer number of new ways to participate that have been conceived in the late 20th and early 21st Century, as well as the old that have been reinvigorated – open innovation challenge prizes, volunteering programmes, mutual aid, participatory budgeting, citizens' assemblies, crowdfunding, to name but a few.

Later, this book will dive deep into concrete examples of charities, businesses, and governments, providing a blueprint for Citizen transformation across sectors. But I want to give a taste here. To show how far the Citizen organisation must extend beyond the usual chatter about sustainable capitalism or ethical business or the future of government, I've selected three examples that represent almost everyone's favourite organisations, our sacred cows. It's true that from within the Consumer Story they represent some of the best on offer. But they could do so much more.

❊❊❊

In the charity sector, consider the concept of effective altruism. This global movement to promote efficiency in philanthropy was popularised by Peter Singer's *The Life You Can Save* and William MacAskill's *80,000 Hours*. A pure application of utilitarian philosophy that focuses on the greatest benefit for the greatest number, effective altruism is based upon randomised control trials following classic scientific methodology. It takes the emotions out of philanthropy and replaces them with clean, rational calculations, including one that estimates the cost of saving a life as between \$2,000-\$3,000.[7]

To its credit, it has spurred people who can afford to give, to give more, by supplying the data of a solid return on the investment. The giving platform that MacAskill created, Giving What We Can, had received \$126 million in donations from 4,571 people, with a further \$1.5 billion pledged over their lifetimes, between its founding in 2009 and the spring of 2020.[8] And that's only one vehicle: many other 'mega-philanthropists' have adopted the effective altruism framework, like the Open Philanthropy Project – created by Mark Zuckerberg's roommate at Harvard, Dustin Moskowitz – which grants \$150-200 million per year.[9] The money goes to projects that can evidence their impact in these clear, measurable terms, like malaria nets and deworming tablets, and which give donors the biggest bang for their buck.

Of course when we approach charity from within the Consumer Story, effective altruism feels like the winning strategy. The tell-tale signs of Consumer logic are everywhere: the focus on data, transactions, return on investment. I'll buy my way to a better world, to a clearer conscience, to heaven – that's the dynamic, the assumption. What is eclipsed by the calculations, however, is the complex context that gives rise to any of these charitable 'causes.' History, trauma, economic exploitation, relationships – all these and other real life factors are missing from the equation. As so often in the Consumer Story, this context is ignored while the choices on offer are accepted as the limits of possibility, as if human life and society were one big series of clearly bounded calculations. Following the logic through, each of these calculations should then be made as coldly and rationally as possible. In the most crass example, when faced with a hypothetical choice of saving a child or a Picasso painting from a burning building, the effective altruist saves the painting, because with the millions of dollars that the painting is worth and could be sold for, many more children's lives could be saved.[10]

Stepping into the Citizen Story, however, we recognise that context not only cannot be ignored, but that context is all. As Citizens, we must develop a sense of belonging in community, cultivate relationships with one another, help heal each other, and collectively build the world that will cease to make philanthropy necessary. The famous quote from Dr Martin Luther King, Jr. is the most apt expression here: 'Philanthropy is commendable, but it must not cause the philanthropist to overlook the circumstances of economic injustice which make philanthropy necessary.'[11] The ideal of the Citizen approach is having everyone pay fair taxes, to a government in which everyone has a meaningful stake; philanthropy explicitly works against this by shielding wealth from taxation. In the Consumer Story, we ask ourselves what we as individuals are getting in return for our taxes. In a Citizen society, we recognise the tax we pay as our contribution to sustaining the whole.

In fact, the Citizen approach entirely rejects the notion of 'altruism,' which derives from the Latin meaning 'other.' Instead, we need to embrace 'proximity' and 'interdependence.' Proximity refers to the truth that, while outsiders can bring insight and understanding, those most affected by a given issue also possess vital wisdom about it. It is therefore these people who must be resourced so they can implement solutions as they see fit. In other words, a transfer not just of money but of power – which is a far cry from the predominance of highly educated white men engaged in effective altruism. Citizens give through participatory grant-making, where decisions about how to allocate money involve not just donors but also recipients, in at least equal representation.

Interdependence is the acknowledgement that we are all connected. This is a basic assumption of many indigenous peoples. No one is truly safe until we are all safe; no one truly thrives until we all thrive. In this view, giving is not regarded as 'charity,' but as 'reciprocity.' The Native American leader and philanthropy professional Edgar Villanueva says it best in his book *Decolonizing Wealth*:

> *Reciprocity is based on our fundamental interconnection – there is no Other, no Us vs. Them, no Haves vs. Have Nots. Reciprocity is the sense that I'm going to give to you because I know you would do the same for me. No one is just a giver or just a taker; we're all both at some point in our lives. This also reflects a cyclical dynamic, as opposed to a one-off, one-way relationship.*[12]

In fact, if the root of altruism is 'other,' the Citizen approach represents nearly its opposite, with the roots of 'Citizen' – as we will see in Chapter 1 – giving it the literal meaning 'together people.' As Citizens, we are meaningful as individuals only through our interconnection. As Villanueva puts it so powerfully, 'there is no other.'

❋❋❋

Next, let's consider the apparel company Patagonia. In the corporate realm, these are the good guys, the paragon sustainable business everyone loves to love. But the Consumer Story still shows up in and constrains Patagonia; stepping into the Citizen Story could take even this company to another level.

Patagonia was born when founder Yvon Chouinard realised that his gear was damaging the rock faces it was his passion to climb, and began making nature-friendly alternatives. The company has pioneered numerous exemplary sustainable practices, like making fleece out of recycled plastic bottles. In 2007, it launched the Footprint Chronicles, which reveal all the suppliers and factories in its supply chains – the materials they use and the conditions for their workers. On Black Friday 2011, Patagonia famously printed a full-page ad in the *New York Times* with the unforgettable headline *DON'T BUY THIS JACKET*, reminding consumers to think twice before buying, a business calling out the impact of consumption on the planet. It even offers lifetime guarantees and free repairs for its apparel, because according to CEO Rose Marcario 'the single best thing we can do for the planet is to keep our stuff in use longer.'[13]

For all Patagonia's unquestionable excellence, there it is, hidden in plain sight: the Consumer Story. The single best thing we can do for the planet is to keep our stuff in use longer? Really? That's the greatest contribution we can make?

The sentiment is echoed in a post entitled 'Introducing the New Footprint Chronicles' on its website: 'You and I have the power to change the habits of our world by changing our buying habits and doing what we can at work to reduce human and environmental cost.'[14]

The thing is, as customers of Patagonia, we're still left in a place where our only real kind of agency is whether we buy or don't buy, or buy from them rather than from someone else.

The only participation on offer is the act of buying, or maybe donating money to one of the environmental charities it recommends. That's the extent of our power. And that is Consumer power, not Citizen power.

Patagonia's ambition is essentially: let's keep humans from harming the planet. Do as little harm as possible. That indicates an assumption that humans are (a) separate from the planet and (b) fundamentally and inevitably damaging. This is the Consumer Story speaking. In this story, the best we can possibly do is to not fuck things up even more. Humans consume and destroy; the best we can aim for is to minimise the destruction. The identity that we must automatically adopt inside this worldview is the guilty role of 'destroyer' or at best the altruistic role of 'saviour,' both of which keep us apart from nature.

As the Citizen Story takes shape, we are finally realising that there is a deep fallacy here, one that is a big part of the problem. We are not external to nature, Consumers of it. We are nature – participants in it, Citizens of it. Indigenous leaders believe that this difference in mindset is the difference between living a harmonious and sustainable life on the planet, versus our current tragic trajectory.

So what would it look like for Patagonia to step into the Citizen Story?

First and most fundamentally, the mission would have to change, the saviour complex thrown off. No company is going to save the planet; the planet does not even need saving. What if Patagonia instead saw its mission as to equip humanity to be nature? How might a company enable and support us to experience ourselves as an inextricable part of the biosphere, and as such, empowered and capable and responsible in a much deeper way? How might Patagonia involve us, beyond just selling us less damaging stuff?

Such an approach would not challenge everything the company does; rather, it would reframe it, and as a result open up so much more. The emphasis would be on maximising connection, inspiration, and participation. For example, the Footprint Chronicles might be refocused to give more space and emphasis to the positive inspiration Patagonia's designers take from nature; problems in the supply chain could lead to open innovation challenge prizes; solutions might be made 'open source,' allowing and indeed encouraging others to copy; the company could offer and promote qualifications in biomimicry, i.e. design inspired by nature.

Marketing might do less to indulge self-hatred ('Do not buy this jacket') and more to embrace a role in inspiring love of nature; I can imagine an 'I am nature' campaign, crowdsourcing tales of Patagonia-equipped deep experiences of the natural world. Governance might give genuine power to the indigenous people who retain the worldview Patagonia would be championing; or indeed to the customers – now seen not just as Consumers but as Citizens, participants, fellow travellers – through an equity offer, or even becoming a cooperative. There are signs that some of this is starting to happen. For instance, the company has just launched a new initiative seeking to support community energy schemes across the world.[15] This is very much a model in which people are participants and Citizens, not just Consumers.

What difference would it make for Patagonia to fully embrace the Citizen Story? Rather than selling products that make people-as-Consumers the gentlest possible destroyers, the company would be equipping people-as-Citizens to be the human embodiment of the beautiful nourishing presence of the Earth that was its founding inspiration. That would be quite something.

Then imagine if every 'purposeful business' were to join them.

❋ ❋ ❋

Finally, turning to the institutions of government, and the level of the nation: consider New Zealand. Who doesn't want to move there? Who doesn't love New Zealand under the leadership of Jacinda Ardern? The island nation is widely held right now as the best we can be, the place from which the rest of the world takes hope and inspiration. Ardern voluntarily formed a governing coalition with other political parties (notably including the Green Party) despite having a massive majority. Her sensitive response to the attack on the mosques in Christchurch rightly gave her hero status, as she donned a headscarf in solidarity with the victims, and passed a ban on assault weapons. The country's response to the Covid pandemic has also been touted as exemplary, with Ardern using Facebook Live chats for Q&A sessions directly with ordinary people, which showcase her approachable, charismatic and compassionate communication style. We all just need to elect more Arderns, right?

With nothing but respect for Ardern... no.

This is no fault of Ardern, but a call to all of us to step into a greater role, and a call to the institutions of government to open that role to us. In the Citizen Story, politics needs all of us, not just the Great Leader standing out in front, however benevolent and of whatever gender and disposition. When we look to the government official we 'purchased' with our vote to make all the bad things go away, to let us off the hook, that's a fundamentally disempowering and anti-democratic arrangement.

As it happens, the Christchurch tragedy's other major outcome was further heightening of a mass surveillance programme in New Zealand, already flagged in 2014 by the whistleblower Edward Snowden as intrusive and total.[16] There are serious academics who complain the pandemic's lockdown worsened the country's 'democratic deficits,' with parliament adjourned and the media restricted. 'The imbalance between those with power, and those trying to hold them to account, is getting worse.'[17] Meanwhile the issue of housing has been a top concern of New Zealanders for at least a decade, but one that the government consistently fails to prioritise. New Zealand has one of the most unaffordable housing markets in the world, with one in every hundred people homeless, while the government's affordable housing programme falls far short of its goals. The indigenous

Māori people and Pasifika people are disproportionately impacted – the issue of their self-determination intersecting with housing. Leilani Farha, the UN special rapporteur on the right to adequate housing, has called New Zealand's situation a human rights crisis, and pointed specific criticism not only at the absence of a capital gains tax (an omission that favours the wealthy, and real estate speculation), but also at the three-year electoral cycle, which keeps governments in campaign mode, aka sales and public relations mode.[18] Aka Consumer mode.

In fact, a dark cloud looms in New Zealand. In recent decades the country has become the real estate market of choice for the superwealthy, a preferred site for their apocalypse bunkers and boltholes. PayPal founder Peter Thiel, the 'techno-libertarian' billionaire backer of Donald Trump, was granted citizenship in return for agreeing to invest in New Zealand start-ups, without having spent more than a fortnight in the country. He purchased his citizenship, and then purchased a 477-acre property.[19] Before a ban in 2018 on foreign buyers brought the mad rush to a halt, dozens of other billionaires from Silicon Valley and elsewhere had likewise secured properties. Pair their nearly unfathomable financial clout with a sizeable far-right populist party, NZ First, that favours nationalism and protectionism, and what does the future of New Zealand's democracy look like? Is this really a nation set fair as an active member of the global community? Or one in grave danger of becoming the international equivalent of a gated community?

A dark cloud looms, but it is not inevitable. New Zealand could step into the Citizen Story. This would require putting into place structures that actively invite involvement from and build trust with all its people. It would mean seizing the mandate that Ardern and her coalition have to open up participation in New Zealand's democracy, and institutionalise those changes such that they cannot be unravelled by a future administration elected when the money adds up and the energy runs out. There would be a particular focus on involving the 16% who are Māori and over 7% who are Pasifika people, whose trust in a government that they understandably see as a legacy of colonisation is fragile at best.

That might look like citizens' assemblies, participatory budgeting, and crowdsourced solutions to the housing crisis and other burning issues. New Zealand is almost uniquely positioned to do this work, not just because of Ardern's majority, but because the country – and indeed the capital city of Wellington, the seat of government – happens to have a large proportion of the world's most innovative developers of the technological tools that support participation, like the members of the Enspiral Collective who have developed world-leading tools like Loomio for collaborative decision-making and Cobudget for participatory budgeting. They are literally right down the street, and yet Ardern and the government have not invited them in (unlike in Taiwan, which has become the most Citizen government in the world, as Chapter 7 will reveal).

❋❋❋

This work constitutes an approach that one of my mentors, the Israeli strategist and political economist Dr Orit Gal, calls 'social acupuncture': finding the intervention points in society where the energy for a new way of doing things is ready to flow, and releasing it.[20] In shifting the Consumer Story, the primary intervention points are the old organisations that tell and retell and retell that Consumer Story through their communication with us, the people. Each time we help an organisation adopt the Citizen Story instead, that releases energy, creating a new point of light for the rest of society to follow. The vital work in this time is to understand that our failing institutions can be reformed and reinvented from the Citizen Story; indeed, that we can reform and reinvent them, because that is what Citizens do.

THE TASK AT HAND

This book is intended to make sure we can embrace an alternative that too often in our history has fallen by the wayside, but represents our deepest selves. The Consumer Story has us in its grip today, and the Subject Story looms in the background. My aim is to make it possible for us to step into the Citizen Story, and to recognise and dismantle the other two.

The foundation for this work will be seeing ourselves anew: we need a deeper understanding of human nature, embracing the truth that we are all creative, capable, caring creatures. Understanding this and reclaiming the language of Citizenship is the task of Part I.

This is who we are, but it is not who we tell ourselves we are.

Part II is about making visible the stories of self and society we live with and within now and in the past, how they filter our view of the world, and as a result obscure the very real possibility of change. I will unpack the Consumer Story that has been the air we've been breathing for nearly 80 years (but no longer) and the Subject Story that came before it. I will explain the way these stories manifest and function; and what happens when they shift and change.

Finally, in Part III, we are ready for the real task, not just of seeing the Citizen Story, but of stepping into it and unleashing our power. Once we see ourselves as Citizens, we demand that our organisations and institutions treat us as such. In this section, my aim is to provide the tools we will need in order to redesign and reinvent every institution and every organisation, across every sector, in the spirit of the Citizen Story. While that might sound overwhelming, and indeed there are no magic solutions, I will show that the work itself is actually more about getting out of our own way and having a lot more fun in the process. In fact, what we are currently engaged upon is far more difficult and exhausting – like pushing water uphill. We stagger from crisis to crisis, struggling to respond from inside a Consumer Story which pits us against each other and indeed against ourselves, even as we most need to care and collaborate. Embracing the Citizen Story not only gives us the chance of a future; the process itself can be joyful, healing and fulfilling.

We need to see ourselves anew; and we need to do it now, before it is too late.

PART I:
SEEING OUR POTENTIAL

The absolute precondition for the Citizen Story is belief in ourselves and in human nature as creative, capable, and caring, rather than lazy, self-interested, and competitive within a zero-sum framework. Any attempt to redesign our institutions will fail if we haven't embraced this fundamental belief.

There's a moment that occurs often in the work of the New Citizenship Project, when a veil drops. It's the moment in which a client sees for the first time how the idea of the Consumer infuses their everyday life, at work and at home, how in so many ways they live and breathe and embody the Consumer Story; and at the same time recognise it doesn't have to be that way. That they could just as easily step into a bigger idea of themselves and those around them – colleagues, customers, suppliers, friends and family, everyone – as Citizens. Because despite the dominance of the Consumer, all the ingredients for Citizenship are right there, innate, waiting for activation. This is who we all are, intrinsically.

This moment comes as a revelation, a word I use despite or perhaps in part because of its religious associations. A revelation heralds more than new knowledge or evidence, more than just a realisation of what is not yet known. When we experience a revelation, our whole perspective shifts; everything changes, because we change. We see with new eyes.

To embrace the Citizen is in some ways a simple shift. When the veil drops, when the scales fall from the eyes, they fall in a moment; the world changes in an instant. Yet acknowledging the Citizen in ourselves is often easier said than done. It bucks the prevailing wisdom that would have us believe that humans are lazy, greedy, self-centred, and apathetic; that we can't be trusted to do anything but mess things up further. It risks being judged naïve or unrealistic. And once embraced, the Citizen also mandates that we become active instead of passive, that we commit rather than complain, that we expand empathy rather than sink into apathy.

To see and believe in our power, our agency, doesn't mean we are anarchists or libertarians who disregard the role of government and other institutions; nor do we hold ourselves to be masters of the universe or knights in shining armour. It is to acknowledge the inherent worth and potential and power of every one of us to contribute what we each uniquely have to offer, just as all the lifeforms in an ecosystem have a vital role to play. To accept that all of us are always smarter than any one of us. To trust ourselves and each other.

In this section I introduce Immy, Bianca, Kennedy, Reen, and Billy as five emblematic Citizens. At first glance they will seem like ordinary, normal people. Nothing special. Indeed, that is the point. While each is doing exceptional work in their communities, their societies, even in the world, they are far from the exception. Every day millions upon millions of small acts of generosity and humanity and art take place all over the world, proving that a different story of ourselves is not just possible, it's most definitely here. When we open our eyes to the number of Citizens who are working to reimagine and rebuild the world, we start to see what's possible, and who we are capable of being.

1. CITIZENS EVERYWHERE

IMMY
Birmingham, England

I don't have to know Imandeep 'Immy' Kaur for long before I become 'bab,' a local term of endearment that translates best somewhere between 'love' and 'mate.' It's not a superficial expression. Immy really cares – for the work she does, and for the people with and for whom she does it – and is not afraid to show it. This might make her uncool, when cool is measured by the telltale shrug of the shoulders and not giving a shit about anyone or anything. Immy's brand of cool is being on fire to help others. She makes me (everyone) care too; her attitude and sheer depth of feeling is seriously infectious.

It's all the more impressive when I consider her environs. We might expect this generosity of spirit in a small town, but Birmingham is a big city, Britain's second biggest by population, the only one other than London with over a million residents. It's industrial, rough and gritty, prey to the same fate as industrial cities across much of the West: as British manufacturing declined in the 1970s and 80s, so did Birmingham. It was left unloved, and, following a classic 'brain drain' dynamic, its brightest talents tended to leave for better prospects in London or elsewhere. 'You were raised to believe Birmingham was a bit shit and you were embarrassed to be from there,' according to Immy.

Much of Immy's commitment stems from her faith. As her turban and her name indicate (all female Sikhs take the name Kaur, meaning 'princess,' just as all male Sikhs bear the name Singh or 'lion'), Immy is a Sikh. Sikhism has at its core the principle of Sarbat da Bhalla, uplifting and ensuring prosperity of all. It teaches rejection of egotism, a compulsion to share, and an obligation of service to others for the benefit of the community.

Sikhism is a new religion, relatively speaking, that sprang up in the 1600s in the Indian state of Punjab out of frustration with the constant warfare between Muslims and Hindus, instead espousing tolerance and peace. On the celebration of Diwali, the Hindu festival of lights, Sikhs remember Bandi Chorr ('prison release'), when in 1619 a Sikh guru was offered freedom from his imprisonment in the Mughal Emperor's fort. The guru agreed, but only upon the condition that the other prisoners – all Hindus, all political prisoners/prisoners of conscience – be released as well. The lesson: we can accept and embrace our individual freedom only with the freedom of others, even when they consider themselves our enemies.

Sikhism has often met with intolerance in India since its emergence. In the aftermath of crackdowns in the 1970s, economic deprivation led to a wave of migration. This brought Immy's parents to Birmingham, where they met. Facing a new onslaught of intolerance as brown people in 20th Century Britain, they began to make a home, and Immy was born.

The quintessential offspring of South Asian immigrants, Immy worked hard and succeeded at school, and was doing the same at university when misfortune struck her family. Her masi – which we translate as aunt but literally means 'like your mother' – was diagnosed with terminal cancer. The family gathered at her bedside for the last weeks of her life. Immy's grief was overwhelming: 'It was the first time I felt real real grief, like you've been punched and you can't breathe or think or be.'

Thrown by the loss, Immy deferred her studies for a year and travelled to the Punjab, becoming a volunteer aid worker. The experience of reconnecting with her roots – the rich heritage of growing food together, caring for and sharing with each other – started a change in her. Until this time, she had thought of her family's intense connection and traditions as slightly shameful, something that made her odd. Now she began to see that these constituted, in her words, 'a superpower.' She enrolled in a new programme back in Britain: a Masters in International Development, with a major fieldwork component. She thought it would be the best of all worlds: a way to continue her community work in the Punjab while gaining a degree, splitting her time between immediate family and deeper roots.

During her studies, Immy was awarded a fellowship with the

Tony Blair Faith Foundation in London. But her disillusionment with the field of international development was growing: for all the positives of the experience, the opportunities and the learning, what became unavoidably apparent to her was the underlying saviour complex, the perpetuation of the colonial dynamics between those who had exploited and those who were dealing with the impacts of that. Immy's next job – working with a housing association in Birmingham that aimed to provide affordable housing – furthered her rude awakening. Again Immy came up against rigid structures that prioritised gains for the people at the top of the organisation's hierarchy as opposed to truly solving the very real – brutally real – issues of rising rents, mercurial landlords, insufficient affordable housing, looming homelessness and constant insecurity. It was a traumatic realisation to understand that these organisations that were supposed to care often did not – could not – actually care.

'People who really care are just not resourced at all,' Immy observes. 'The people close enough to act and really provide support aren't given the opportunity to affect or even see the bigger systems they're working in.'

Rather than give up, however, the embers inside her were stoked to a furious flame.

In 2011, Birmingham joined the ranks of cities worldwide offering a TEDx conference – the distributive model that allows volunteers to apply for a free licence to host a local offshoot of the famous platform for 'ideas worth spreading.' An ostensibly 'Technology Entertainment Design'-focused conference might not seem like something an aid worker with a social justice focus would embrace, but after a chance meeting with Anneka Deva, the young activist who organised that first event, Immy was intrigued and signed up as a volunteer. Her pride and passion for Birmingham – Brum, as it's affectionately termed, hence TEDxBrum – blossomed with the experience. 'I learned for the first time,' she reflects, 'that you can just make what you want to see. That people can do that.'

Soon she would take over the role of Curator (still a volunteer position, if a hefty responsibility, which she accomplished alongside her day job at the housing association). The event was even luring home a handful of Brummie natives who'd left to chase the bright lights in London, a phenomenon elsewhere called 'boomerangs'[21] – and they joined the team too. The line-up she and the team put together in 2013 under the theme of 'Marking the Map' included a rich mix of spoken word artists, community organisers, educators, and designers.

One of those in the audience at TEDxBrum in 2013 was Indy Johar, a renegade architect and thinker, and fellow Sikh, 11 years Immy's senior. Indy had founded an organisation called Dark Matter Labs as an architecture-practice-meets-think-tank-meets-innovation-consultancy (my summary) when he realised that the kind of housing projects he wanted to develop – projects that prized community and human connection – were virtually impossible within the framework of contemporary property rights. This is an example of the archetypal 'dark matter' that the organisation's name invokes: the structures and assumptions which sit below the surface and shape what is possible; the need for architecture of human systems not just physical buildings.

Indy saw that the organisers of TEDxBrum were onto something, and he made sure not to leave without finding Immy. They became instant friends. Soon after, in typically direct fashion, he confronted her. '"What are you wasting your time for, doing this job you don't believe in, working within all those restrictions?" he said,' Immy remembers. 'He was telling me, like ordering me "You need to be doing this work full-time. This is what you're here for, you know that." He didn't know what it meant, or have anything specific. I didn't either.'

What she did know was that, as much as she was enjoying TEDxBrum, a one-day event wasn't sufficient to revitalise and sustain her beloved city. 'We soon realised that ideas weren't enough, one day a year wasn't enough, volunteering alone wasn't enough, and not having a sustainable revenue generation through TEDx events, would eventually mean the energy would run out as other pressures were faced by the team.... It was then we realised it was crucial to build something more long term, a more permanent way to showcase possibility, build, create, prototype and scale solutions as we worked on the pressing challenges facing our city.'

Inspired by Indy, Immy set out on a new mission, to bring another kind of franchise to Birmingham, but this time one that would achieve a constant, ongoing presence: Impact Hub Birmingham. Impact Hubs, at a nuts-and-bolts level, are co-working spaces for social entrepreneurs. They pioneered the idea of hosting, where rather than simply offering that standard co-working benefit of 'resource efficiency' – sharing a receptionist, copy machine, and coffee machine – individuals and organisations using these spaces would be provided with opportunities to come together, socially and professionally, to learn from and spark off one another's work. 'Serendipity engines,' per Immy and Indy. Not unlike TEDx conferences, the organisation runs on a distributed model whereby – after a peer review process – almost anyone can establish one.

Immy wanted the Hub not just to gather organisations that wanted to make the world a better place; she intended to convene those who would make it actually happen, right there in Birmingham, to change the city. This would be articulated to funders and supporters as Mission Birmingham; the founders and the community refer to it as #EpicBrum.

The team came together quickly, many of them the same crew that had delivered TEDxBrum and shared Immy's desire to do more. Then came the space, and the crowdfunder. With no real indication they could fundraise the amount required, the team committed to a five year lease on a warehouse a ten minute walk from the city centre.

When I went to visit for the first time three years on, in the middle of 2018, the place was buzzing. A wall of mugs greeted me on entry, performing a dual function as an in/out board – if someone's in the space, they take their mug, and when they go, they wash and hang it up. Coffee played a crucial role, with a barista-standard machine open for all to use, with regular training sessions to equip newcomers. It felt like something between a high school tech lab and an office, with self-made desks, a 3D printer, and graphic recordings of recent events all over the walls. Outside, too, I could sense the impact the Hub had made on the surrounding area. Cafes and bars were opening amidst the shells of bingo halls and weed-strewn car parks.

This do-it-yourself vibe clearly set the tone for the #EpicBrum work that took place in the space: among over a hundred individual members, each working on their own initiatives to better the city, were some real showstopping collaborations. Democratising Development, or #DemoDev, has won several major awards for opening up local authority data to help identify spaces for small-scale, citizen-led, decentralised housebuilding – rather than any development requiring big spaces and major companies. Birmingham is now something of an outlier in a country with a major housing shortage, a success to which #DemoDev has made a significant contribution. The Radical Childcare project, which was born out of the idea of one member to create a shared, on-site creche for the community, has grown into a parent-led and -developed policy campaign, and spawned similar experiments across the country.

Yet as the end of their five-year lease approached, the landlords, who had never really engaged with the mission, decided to increase the rent significantly – largely owing to the success and vitality that the Hub itself had brought to the immediate area. Immy and the team knew they were finished. It was a setback, and it hurt. But within weeks, Immy turned her attention to what would come next.

Conventional logic might have taken them to a bigger space, even closer to the heart of Birmingham; or perhaps even to another city. Scale the model, or spread it. But no. That's not where their learning took the team. 'We didn't want to go bigger and broader. We actually felt that it was when we'd got tight in, worked with the problems people were actually facing for real, right in front of their faces, that we'd opened up the really big change. So no, we didn't want to go bigger, or at least not broader. Not city. Neighbourhood.'

Having earlier gone from development work in the Punjab to catalysing community in Birmingham, Immy likes to quote adrienne maree brown, the author of *Emergent Strategy*, who contrasts 'mile wide, inch deep movements with inch wide, mile deep movements that schism the existing paradigm.'[22]

This was the genesis of CIVIC SQUARE. First established as a challenge to the developers regenerating a Birmingham neighbourhood called Port Loop, it has since become a partnership with them. Lying about a mile to the west of the city centre, the heart of Port Loop is an island surrounded on all sides by the canals that meet in the centre of Birmingham, one of which 'loops' around to rejoin the other, originally to serve a cluster of warehouses and factories that sprang up in the city's late 18th Century industrial heyday. The ambition for CIVIC SQUARE is somewhere between Impact Hub, library, community centre and Roman forum: a place where local people will come together on an ongoing basis to invent and reinvent the place they live; where they will be equipped to own and lead their own community, rather than be told what to do.

At the time of writing it's just beginning, with the first iteration hosted on a barge, dubbed the 'Floating Front Room.' Locals are treated as neighbours coming into an extension of their own homes, offered coffee and cake, conversation menus to get them talking, and bike repair and maintenance workshops. The idea is to start small, at the scale of the new neighbourhood, and grow with it. In tandem with this inch-wide starting point, Immy simultaneously established the mile-deep aspect, in the form of the Re Festival. This six-day online ideas festival was held in the first weeks of the pandemic, and saw an array of the world's most original thinkers and doers invited to share their ideas and insights at the very beginning of CIVIC SQUARE. This kicked off a strand of the project called the Department of Dreams, a set of resources that is constantly evolving; it provides the work with a ready supply of inspiration on tap, and makes that inspiration available to any other individual or initiative that might seek it out.

For Immy, this narrowing and deepening of focus feels like a sharpening she was born for, an arrival in the work she was meant to do.

She writes:

I get asked a lot in Birmingham, do you really believe that system change can happen, and well my answer is yes. When we talk about 'the system,' to me that is a set of entrenched powers and structures that exists, these systems exist to sustain themselves, and those who run them or benefit most from them wouldn't benefit from them ever changing, why change it, it works well for them....

The thing is, I still believe that if we want something to change, we need to build alternatives, show a different way of life. We are the people who are going to do this, the role models, we are the real rock stars, not those in power, not the media and not celebrities. We can't simply throw our arms up in resignation and sit back, or wait for someone else to come and lead, we kind of have to do it for ourselves, and we kind of have to get on with it now. This last year has shown me this passion, this fire, this will to create a more equitable, democratic and imaginative world exists, it exists right here in our home city and it is alive and well. It's all on us, to do it together.[23]

BIANCA
Berlin, Germany

It's hard to articulate the particular dilemma of living in a place that is functional and stable relative to others, but at the same time still riven by all the same contradictions. If and when someone speaks out about these, as Bianca Praetorius often does, and is met with blank stares or indignation, as she often is, we can't help recall the story of Cassandra in Greek mythology, who bore the curse of never being believed, despite speaking the truth.

Bianca was born in the mid-1980s in Frankfurt, which along-side its identity as the continent's financial powerhouse – home to not just Deutsche Bank but also the European Central Bank – is populated by a majority of residents with non-German backgrounds (51.2% in 2015).[24] This includes both high-income international finance professionals from Europe, Asia, North America, and elsewhere, as well as lower-income immigrants from Turkey, Afghanistan, Sri Lanka, and other places. At the primary school Bianca attended, she was one of only four white-skinned German kids in her class. Her German still carries the inflections of the creole-German spoken – often with pride, at least in Berlin – by first and second-generation immigrant communities, which sets her apart even when her long blonde hair, pale skin and light eyes make her seem classically Germanic.

Bianca didn't enjoy school. She found reading onerous, and was easily distracted. 'Spending seven hours per day in front of the television – not unusual for a lower-class childhood – zapping through imported American TV shows, this definitely affected

my attention span.' She was impatient with the abstracted information in the classroom, only interested (passionately interested,
in fact) in its applications to the real world, which – tragically –
weren't forthcoming in her lessons. 'Kids like me fail in our school
system, and that's what I did.'

When she later discovered theatre, she liked that it empha-
sised embodiment of the mechanics of the world, as opposed
to written accounts. And so Bianca studied acting and became
an actress. Or, she attempted it. After a series of trips around
the audition-rejection cycle, she pivoted and found gainful
employment inside Berlin's burgeoning start-up scene, coach-
ing fledgeling tech entrepreneurs on how to pitch to investors.
Against the odds, it seemed that she had found her place: the
work was plentiful, and lucrative, and she excelled, her acting
skills finally put to good use.

But then, from this place of privilege, Bianca started noticing
things. One was that many of the ideas coming from the start-up
scene were constrained by German policies. Some of the obstacles were societal – the extremely risk-averse investment culture,
for example – and others were actual laws. It seemed terribly
unfair to Bianca that brilliant innovators elsewhere were actually supported, and were, as a result, able to build future-defining
projects, while in Germany they were often withering on the vine.
Bianca knew nothing about how to change those legal and cultural and structural obstacles, but she started investigating them in
her free time.

The rise of the Pirate Party in Sweden from 2006 onward,
which then launched in Germany and made its way into four

state parliaments (including Berlin) by 2011, offered a fascinating precedent to 'hacking government.' In 2013 Rick Falkvinge, the Pirates founder, released his book *Swarmwise: the Tactical Manual to Changing the World* (for free, via a Creative Commons license, part of the Open Source movement), which immediately acquired cult status among the 'digerati' of Europe. According to Falkvinge: 'everybody can change the world if he or she is passionate about a specific change, and that change is tangible, credible, inclusive, and epic enough to attract a swarm'[25] which he defined as 'a decentralized, collaborative effort of volunteers that looks like a hierarchical, traditional organization from the outside. It is built by a small core of people that construct a scaffolding of go-to people, enabling a large number of volunteers to cooperate on a common goal in quantities of people not possible before the net was available.'[26]

Then in 2016, while mulling over these challenges and opportunities at an organic farm in Brazil (a playground for 'wonderful New Agey self-development stuff'), Bianca saw the headlines about the resurgence of neo-Nazis in response to the influx of Syrian refugees into Germany. Tempting as staying on in Brazil was, 'my house was burning back home.'

Bianca returned home to Berlin straight away to Do Something.

In November 2016, she received the news of Donald Trump's victory in the US elections with the same incredulous horror as most of her countrypeople. The cover of the 12th November issue of *Der Spiegel*, the largest news magazine in Germany (and Europe), depicted Donald's head as a fiery meteor on a collision course with the planet.

There was, however, one group of Germans who outright celebrated the news: members of the far-right party Alternative for Deutschland (AfD). They were already occupying seats in ten of Germany's 16 state parliaments, up from a single seat in 2011, a rise unparalleled in its speed in comparison to other far-right parties in Europe like the National Front in France, and fuelled by the same nationalist, anti-immigrant rhetoric. 'Islam does not belong in Germany,'[27] their platform blatantly announced, their posters featuring two nubile white women strolling along the beach, emblazoned with 'Burkas? We prefer bikinis.'[28]

The AfD would go on to win a major victory in the September 2017 elections, coming in third and entering the national parliament (Bundestag) for the first time ever by winning 13.3% of the votes, which translates to 94 (of 709 total) seats. Many of the AfD supporters expressed their votes as protests against Chancellor Angela Merkel's 'Welcome politics,' which had opened its arms to more than 1.3 million migrants and refugees beginning in 2015.

Convinced that the resurgence of nationalism and hatred could be countered by mobilising more of her generation – among whom apathy and even outright contempt for politics in Germany was common – Bianca hit the ground running in late 2016. Spontaneity, which some (especially in Germany) might call impulsivity, is one of her chief qualities. In a society known for being risk-averse, Bianca is unafraid of failure, which she attributes to her experiences at school and as an aspiring actress as much as to the mantra of start-up culture – 'fail fast, fail often.'

She put out feelers to find out who was taking a fresh approach to politics. A group was just forming, and Bianca became one of the founding members. 'The first question we asked ourselves was, imagine that political parties were invented today: how would we invent them?' Germany's government was explicitly engineered to resist change and innovation in the wake of World War II, with powers distributed between the state governments and the central government, and the constitution barring radical upstarts to prevent another dictator from ever seizing power again. Yet its very stability, predictability, and – it must be said – patriarchy, is now resulting in widespread contempt for politics in younger generations, who weren't alive during the war.

'The way political parties in Germany – and most places – function is that you feel attracted to a party's platform, you meet in real life in the local party office, and you become a member. But our lives today are very mobile – many of us don't stay in one place for very long.' The majority of Bianca's peers – Europeans who grew up in the 1990s, after the Schengen Convention dissolved border controls in the EU – have lived, studied, worked across national borders, in one of the other-member states.

The upsides to this generation's fluidity are many; however, a desire to engage in politics is usually not among them. With a diminished sense of roots, of national belonging, Bianca saw that her peers were even less incentivised to get involved in traditional parties – but also that a deep need and equally deep potential for involvement in something remained intact. 'At the same time, people want to get involved, because we haven't quite forgotten that we are part of the fabric of community and democracy and that we have the potential to get involved.'

Bianca and her new colleagues also perceived a further challenge, in the form of the tension between the value of the individual and the concept of joining. 'The mindset of the last few decades of capitalism telling us that we create our individual identities through the choices that we make in the marketplace is the opposite energy from 'I'm joining a party." Especially when there are only two, or even four (as in Germany), main choices of parties. '[For young people,] the whole idea of being a party member is distasteful – it's almost shameful. That's enough to destroy the whole party system and democracy right there – it currently only attracts people who don't believe in the value of the individual – which is the minority by far,' says Bianca.

That groupthink flavour is amplified by the fact that political parties rely on a powerful few with conventional expertise to draft policies, which the rest of the party members can only approve or reject. Those 'experts' tend to represent a homogeneous group: overwhelmingly male, all with the same academic background, and a lot of ego and righteousness around their deliverables. Thinking outside the box, embracing new technologies, welcoming diverse contributions: none of this is part of their mindset. It furthers the alienation of young people who are accustomed to openness and sharing.

Under the current system, approval of a party's policy platform happens during in-person gatherings with highly-formalised debate processes, with a series of presentations between the 'pro' and 'con' spokespeople. 'It's a combative culture,' says Bianca. 'It creates the opposite of a collaborative approach. It also gets really predictable and boring: certain words and concepts are like triggers that always get the same reaction from certain people.'

There had to be ways to make the process and the experience of politics more inclusive, more engaging, more collaborative. More fun. To capture this spirit, Bianca and her co-founders settled on the name Demokratie in Bewegung (Democracy in Motion) for their new party. DiB built a platform that enabled 'swarm intelligence' to crowdsource drafts of its policies. 'Anyone anywhere could create an initiative that could end up being part of the party's official platform.'

Even more importantly, they introduced rules to shift the culture:

We established quotas: 50% women and 25% minorities (which we defined as anyone who had a discrimination background, whether because of disability or ethnicity or any other kind of discrimination). Each policy idea on the platform had to be co-written following these quotas, and the decisions to finalise and adopt the policy also had to be made by people according to the quotas.

We also introduced a simple hack: a communications rule amongst the members that the conversation had to alternate between a woman and a man every turn. There could not be two women speaking one after the other, or two men. And if a woman did not come forward after a man had spoken, the list was closed. No one could make further comments. Suddenly the men found themselves forced to ask women to speak up, if they wanted to speak. It was a powerful enlightenment moment for them.

As 2017 dawned, bringing with it national elections for parliament in Germany, there were 2,500 people actively involved in creating DiB policies on the platform. Given the way such digital communities tend to be structured – as first noted by New York University's Clay Shirky in his 2008 book *Here Comes Everybody*, in findings widely endorsed since – this suggests the group had built a support base of anything up to 250,000,[29] with virtually no media spend whatsoever. In the end, however, the nearly 61,000 votes DiB received weren't sufficient to land them any seats in Parliament.

Yet the challenge that DiB represented to the old-fashioned, male-dominated, 'expert'-driven process didn't go unnoticed. Among those who took note of the fledgeling party's cultural 'hack' was former Greek finance minister Yanis Varoufakis, who was spearheading a pan-European grassroots movement called Diem25 (which stands for Democracy in Europe Movement 2025). Diem25 maintains that if democracy in Europe isn't fixed by 2025, such that all member states feel equally dignified, the EU will fall apart.

And so it was that Bianca – high school dropout, failed actress, and start-up pitch coach – found herself on the 2018 ballot for a seat in the European Parliament, against all the odds. Bianca tells me that never in her wildest dreams would she have stepped forward and suggested herself as a candidate for the European Parliament. But she was the one who received the votes to rank the second-highest of all potential women candidates in the new joint-party, Democracy in Europe. 'I had to work through feelings about not being experienced enough, my fear of disappointing people. But my tech experience saved me – none of the other people on the list had any, and I felt like I could maybe bring the tech community into more engagement in politics.'

Still, as the campaign got more publicity, the attacks started. Her German was criticised for sounding rough and uneducated. Her background in acting made her 'vain and superficial.' Her work as a coach meant 'she hadn't succeeded at anything real.' Her openness about the failures she had encountered along her path was met with the unforgiving German reaction to failure of any kind: 'She failed at everything else and now she's trying politics.'

The policy platform of Democracy in Europe was a breath of fresh air. It included the Green New Deal for Europe (thanks to Bianca's passion), and a call for a new Constitution for Europe that would be developed in citizen assemblies, to replace the current 'set of treaties drafted by unelected diplomats and government ministers.' It aimed to 'reform Europe's competition laws in order to allow local governments to resist privatization and more easily re-municipalize public services' and to 'change public procurement laws that currently restrict the possibilities for municipalities to promote a radical transition to sustainable and ethical economies' and to 'reform EU asylum law to enable municipal governments to welcome migrants against more restrictive national laws' as part of its mission to end 'Fortress Europe' and instead introduce safe, legal, and open pathways for migrants into Europe.[30]

But when the elections came around in May 2019, DiB couldn't compete with the big parties in capturing attention. 'It was very disheartening to see so much potential, and these policies that would have disrupted Europe in such a positive way, and no one cared,' said Bianca. For a time she lost her trademark boundless energy and grieved.

At the time of writing, she is several months into a new strategy, taking on the challenge of working for change from within a traditional, established institution. To the horror of many of her politically-progressive peers, Bianca has joined the governing Christian Democratic Union (CDU), 'as a kind of Trojan Horse for climate policy.' From there she initiated the KlimaUnion, a group working to develop a climate agenda that goes with the grain of the core conservative values of the party.

It remains to be seen just when and how her relentless innovator's spirit will prevail but there is no question that actress-turned-startup-coach Bianca has become a serious figure in the cold, hard landscape of German national politics.

KENNEDY
Kibera, Kenya

In Kibera, a slum on the edges of Nairobi, between 600,000 to one million people live without running water or (legal) power. It's hard to keep track of how many, exactly.

Corrugated metal sheet shacks huddle along muddy dirt paths by the thousands, with messy and ingenious workarounds to the lack of sanitation and electricity. There are no government schools or health services. It was here that Kennedy Odede grew up, in the late 1980s.

His rebellious mother had refused – escaped from – three arranged marriages, until she found herself pregnant with Kennedy and still unwed. For the sake of her unborn child she accepted a husband so notorious for his violence that he'd found no other wife, because in the village where she came from, a child with no father was liable to be forcibly taken from her and left in the wilderness to die. After he was born, the family moved to Kibera, its proximity to Nairobi promising more opportunity than the village.

They had almost nothing. Almost nothing to eat, many times nothing but oil rubbed on their lips, their stomachs rumbling all night. Almost nothing to wear – going without shoes for years. Kennedy taught himself to read from scraps of newspapers, and then with hand-me-down books from a friend who could afford school. Kennedy's mother was enterprising and banded together with other women in Kibera so they could help each other survive, but every time she scraped together a few pennies,

Kennedy's stepfather absconded with it, usually to drink it away and return full of rage.

In 1994, when Kennedy was 10 years old, unwilling to bear his stepfather's vicious beatings any longer, he ran away and became a street kid. He was taught how to survive on the streets by his best friend, a 13-year-old boy named Kamau. Kamau was stoned to death by an angry crowd. To get off the streets and into a school run by missionaries, Kennedy endured molestation by the head priest – until he could no longer stomach it. A man raped Kennedy's sister Jackie when she was just 15 years old; she became pregnant. After the baby was born, out of options, she moved in with the man who had raped her. Kennedy lost more friends to random violence, disease, or to desperation – suicide. He did hard labour in unsafe factories for pennies, if he was paid at all.

'The life we lived in Kibera was a life of no choices, no options. All we could do was try to survive, and often barely that,' Kennedy recalls. The fact that what everyone first notices and most remembers about Kennedy's appearance is the generous dazzling smile that more often than not overtakes the whole of his round face is a testament to his resilience.

Books were Kennedy's only escape from poverty. American aid workers providing schooling in the slum gave him copies of the work of Marcus Garvey and Dr Martin Luther King, Jr., whose words struck Kennedy deeply: 'Anybody can serve. You don't have to have a college degree to serve. You don't have to make your subject and verb agree to serve. You only need a heart full of grace. A soul generated by love.'[31] He would make them the words he lived by.

When he was 20 years old, unable to bear the way things were any longer, Kennedy gathered his closest friends and held a meeting. He proposed they take matters into their own hands: sanitation, education about women's rights, a soccer team to get the boys doing something wholesome. One friend asked where the money would come from, whether Kennedy had found white people to support these efforts.

'I felt the blood rise in me. For too long my community had been told that we could not do anything by ourselves without money from the outside, without the financial support and wisdom from the Western world. "We don't need money to clean our streets. We don't need money to remove the garbage. We don't need money to protect our little sisters from being sexually abused and raped. We don't need money to play soccer and just be there for each other", he told them.

Kennedy was determined to alleviate the suffering in his community without turning to foreign organisations for help. 'Since we were the ones who understood all the challenges we faced, we also had the best shot at finding the solutions. As my mother always said: only he who wears the shoe knows how it pinches.' Kennedy knew what they were going to call themselves, too: Shining Hope for Communities. When his friends objected that it was too much of a mouthful, they settled on SHOFCO, for short.

Each took up positions of responsibility to begin to make Kennedy's utterly fantastic idea into reality. The core team of seven organised a few dozen others to do regular cleanups in Kibera. They formed a soccer team. They started performing street theatre to show scenes of domestic abuse and rape and what alternatives could look like. Kennedy helped start several hundred small businesses – like vegetable or water stands, or barber shops – with a 'pay-it-forward' mechanism, giving out very small loans from his factory earnings, and then requiring that instead of paying the loan back, the recipient find someone to pass it along to. Every time a new department of SHOFCO was launched, the core group voted on who would be its leader and that person would not only be formally referred to by a title like 'Director' or 'Chair,' they would also have decision-making power. Kennedy was clear that everyone needed to feel like SHOFCO belonged to them, not just him, in order for it to succeed.

Kennedy was invited to speak at the World Social Forum, an international gathering of activists. The speech he gave there drew the attention of people from around the world, but Kennedy was determined to keep the organisation self-reliant, inspired by Marcus Garvey's insistence that Africans must achieve financial independence. It was Garvey who had originally spoken the words that Bob Marley would make famous: 'We are going to emancipate ourselves from mental slavery because whilst others might free the body, none but ourselves can free the mind.'[32]

Among the people of Kibera, Kennedy became known as the 'Mayor' – 'not elected Mayor, it's a name of affection for a man there for his people,' he notes. 'A real Mayor, who takes responsibility for what the politicians leave behind.' Every night a line of people from the community formed outside Kennedy's home, seeking his counsel about their marriage or about conflicts among friends or family members, a sort of ad-hoc justice system. 'A place like Kibera has an unofficial government since the official government couldn't care less,' says Kennedy.

Although those with official power might not have cared less about the lives of the people of Kibera, they took note of Kennedy and how much sway he held with the not-insignificant number of voters in the slums. Kenya's general election of 2007 was much anticipated, pitting the incumbent President Mwai Kibaki against the opposition leader Raila Odinga. President Kibaki was a Kikuyu, the traditionally dominant ethnic group in Kenya, while his opponent Odinga was a member of the Luo ethnic group – like Kennedy – and had united many other tribes, calling himself the 'people's president.' By September opinion polls showed Odinga distinctly in the lead, with Kibera a stronghold of the opposition.

Not long before the December elections, the President's people reached out to Kennedy and offered him a job. They would pay him more in a day than he earned in a month, if he would help them commit voter fraud and rig the elections. All Kennedy had to do was get the names and identification numbers of opposition supporters and bring them back to the office, where the polling places of those voters would be changed without their knowledge, so that when they showed up on Election Day, they'd be told they weren't in the right place. They were not the kind of potential employers who would take no for an answer, but Kennedy evaded them by saying he was leaving town.

On 27 December 2007, the sitting President won re-election, despite much of the world suspecting fraud. The long-simmering tribal tensions exploded, with non-violent protests against the results quickly spinning into all-out tribal warfare, each side clubbing and killing the members of the other – their neighbours until that moment – and the police killing indiscriminately in an attempt to control the chaos.

Because of his status as the 'Mayor,' Kennedy was a particular target of the Mungiki, a notoriously brutal Kikuyu gang. 'All the shops in the neighbourhood have been closed or looted. The road going into Kibera has been shut by the mobs and men in uniforms – paramilitary police. Nothing and nobody comes in and out without a struggle. They are sealing us in to die,' Kennedy wrote later in his memoir.

I feel deeply sad, not only for the victims but also for these young men who have been so badly oppressed that now they have turned wild. Many residents of Kibera simply cannot imagine that they have anything more to lose. I always heard about terrible things like this happening in other countries nearby – Rwanda, Congo – and I thought it would take so much to get to that point. Now I see that it doesn't take much at all, simply a spark, if there is enough poverty and hopelessness to serve as kindling, the flame burns and burns.[33]

A number of Kennedy's friends and acquaintances were killed, but with help from the many people who respected him, he managed to escape through the checkpoints – his fluency in both Kikuyu and Luo also coming in handy – and crossed into Tanzania. The post-election violence continued for several months, with the injured and dying in Kibera receiving no medical assistance or food. Finally in February 2008, with the intervention of the UN, the two parties negotiated a power-sharing agreement and formed a coalition government. Despite ongoing sporadic violence, in March Kennedy returned to Kibera. The buildings and businesses that had been burnt to the ground or reduced to rubble could slowly be rebuilt; what was harder to rebuild was the trust between neighbours who had turned on each other and now had to carry on living side by side.

In the bleak aftermath of the post-election violence, Kennedy was offered a life-changing opportunity: to study and earn a degree at Wesleyan University in the USA. Many in this fortunate position – most – would have turned their back on the slum and never looked back.

Not Kennedy. He stayed in touch with his SHOFCO colleagues throughout, using his time at the American institution to forge alliances, to spread the word about life in Kibera, and to lay the groundwork for what had been his biggest dream all along: a school for the girls of Kibera, with free tuition, run by teachers and administrators who came from the community. And when he graduated from college, he returned to Kenya to make it happen.

In 2009, he and his partner Jessica opened the Kibera School for Girls, providing free education from kindergarten through to the age of 14. From there SHOFCO expanded to offer a free health clinic, a small-business incubator for women with HIV called SWEP (for SHOFCO's Women's Empowerment Program), a safe house for survivors of violence, and clean water provision, which pioneered an aerial system of pipes that prevent tampering and contamination. By 2020, SHOFCO had expanded into 25 slums across Kenya, reaching 2.4 million urban slum dwellers. All the programmes continue to be directed, organised and run by local community members, although international donations and funding have been integrated into the model.

The pandemic could have ripped through the slums, where people cannot social distance and where 50 families often share a single toilet. But SHOFCO created handwashing stations, diverted the women of SWEP to making and distributing soap and hand sanitizer, created a campaign to combat disinformation about the virus, and provided meals for students once schools had to be closed. Though not a disaster response organisation, SHOFCO was primed to respond immediately because of its proximity to the community. Kennedy has been credited with averting a catastrophe.

He is more committed than ever to growing and supporting fellow citizens in Africa so they can create their own solutions, and overturn the colonial legacy of international aid and foreign philanthropy. He writes: 'In order for marginalised communities to break out of survival mode, the entire community must transform at a systemic level, driven by economic opportunity, access to basic services, and a belief in the dignity and self-worth of every individual.'[34]

REEN
London, England

Irenie 'Reen' Ekkeshis and I first became friends in 2006, when we worked together at a London advertising agency. We were an 'account team': the division of labour made her the team manager, in charge of the client relationship, and me the strategic thinker, the researcher.

It was an artificial division, since she is deeply thoughtful and I am occasionally capable of looking after myself, but one that retained some truth and use throughout our work together. Reen is a consummate communicator, with an uncanny ability to understand almost everyone she meets better than they do themselves. Her belief is 'if communication doesn't work, it's not good communication': the responsibility lies with the communicator, not the audience.

I missed her deeply when she left in late 2007 to take over as the young CEO of a cultural heritage tour operator, working with mostly male, older, highly expert museum curators. The company would soon be bought by a bigger player in the market, with her path upward clearly laid out.

As her name indicates, Reen is of Greek descent - Greek Cypriot to be exact. Her favourite wedding photo shows her creased in laughter on the dance floor, her Liverpudlian bridegroom Al having just performed a full Greek routine secretly taught to him by her brothers.

A year on from the wedding, in January 2011, Reen woke up with a streaming, itchy eye, as if there were a piece of grit in it. She wore glasses that day instead of her usual contact lenses, but by the evening photophobia had set in, the kitchen lights aggravating a searing headache. Next day, she went to Moorfields Eye Hospital, a short train ride into town, but was met with little more than confusion and eye drops. She went on to work. But by the end of the week, she could barely move for the pain in her eye and head.

She went from one cornea specialist to another, and another, eight or nine in total, each asking the same questions in the same manner, 'kind and compassionate... but also really aggressive,' trying to identify what she had done wrong, find a cause in her behaviour. Had she done x, y, or z? The professional medical inquiry felt more like an interrogation. It would become familiar in the months and then years that followed. The final diagnosis was a rare condition: a kind of amoeba called the acanthamoeba. They live in soil but also in all kinds of water: tap water, swimming pools, and so on. We drink and eat them with no ill effect. But if held against the eye by a contact lens, they can transfer to the cornea, the transparent shield layer of the eye, and become deeply destructive.

The initial treatment regime was frightening enough: unlicensed disinfectants – polyhexanide, more commonly used to clean swimming pools – would be diluted and applied to her eye literally every hour for at least three months. But given precious little information as to what to expect, Reen did what we all do and turned to Doctors Google and Facebook.

What she found terrified her: concerns (ultimately disproven) that the amoeba could migrate down the optic nerve to the brain, with potentially fatal consequences; experimental and uncertain treatments; one poor woman who was in such horrific pain that she had had brain surgery, which had in turn triggered a stroke.

By comparison at least, Reen was lucky. One of the world's leading eye hospitals happened to be a short, direct train ride from home, and she was soon having weekly appointments. She had her mother by her side at all times, Al and her dad too when possible. During the appointments she and her mum left the car under the watch of a Turkish family who had known Reen's mother since childhood, who ran a chip shop and dodged the parking tickets for them, and provided free-flowing chips after each visit. Reen devoted the next 15 months to excruciating treatments, not just for the acanthamoeba itself but for the many complications that arose. Finally, things stopped getting worse.

Reen had no sight in the eye, her face was badly swollen and she was still in constant pain. She had lost her job, her confidence was battered, but she began contemplating what needed to be done so that other people wouldn't suffer as she had. She suspected the problem was communication. Reen knew she was not to blame for the suffering she'd endured. What was the right thing to do next? To launch a court case against the hospital, or the contact lens industry? She was advised that the only route to any compensation was probably to sue the hospital, but she did not want to do that.

The many conversations she'd had since her ordeals began had made one thing apparent: no one really knew that acanthamoeba live in water and are a real danger to anyone who wears lenses. As she contemplated the situation, she realised that there were two intervention points. First, and most immediately, a visual warning on the packaging of contact lenses, so that lens wearers were immediately informed of the danger of contact with water. Second, remembering the onslaught of interrogation and accu-sation she had faced, there was something about the web of re-lationships between producers of healthcare, hospitals, doctors, and patients that needed to change.

With her husband, a talented art director, Reen designed a warning graphic: the universal language of a tap with a drop of water, circled and crossed through in red. On Twitter, she con-tacted the British Contact Lens Association (BCLA), getting an email address and a response that left the door ajar, after which there was no stopping her. When the trail went cold and she was pushed away, she traced the President of the BCLA, called her, and asked for an invitation to the upcoming annual lecture, which happened to be on microbial keratitis. Once there, Reen made the President her friend and ally, and soon enough the BCLA agreed to create the icon as stickers and promote them to opticians.

Meanwhile, in partnership with one of the team at the hospi-tal, Moorfields, Reen co-authored a research paper, published in a major journal, which finally identified contact with water as a major risk for contact lens wearers. It included a section on the implications and considerations for marketing communications in the sector. The Australian and American equivalents of the BCLA went one step further than the British, and began provid-ing stickers for free.

Back at Moorfields, Reen's eye drops and painkillers contin-ued but at lower concentrations, and the weekly visits became calmer – but in a way worse. Moorfields is a teaching hospital, and interest in Reen's condition was high. She was studied, asked the same questions again and again, asked permission but then largely treated as an object. But then the doctors heard about her No Water work. They started to look at her differently, to see her as a human being with an eye condition, with an expertise and a contribution that could stand alongside and complement their own, not just as an eye condition attached to a human body.

With this new-found support, Reen turned her attention to the eye research charity, Fight For Sight. She made contact, citing the lack of research on her condition and those associated with it, remembering the papers Dr Google had offered her. The Director of Research sent her an abrupt note on the rarity of the condition and the abundance of other, more pressing priorities. Of course Reen would not let it end there, and transformed this relationship too into an alliance. Soon, she had an invitation to an experimental process called a Priority Setting Partnership, bringing together researchers, clinicians, patients, and funders from across the field of ocular medicine, to set Fight For Sight's research priorities.

This group met in April 2013. The participants discussed and debated, made trade-offs, and eventually delivered a set of priorities far different from anything that would have come from researchers alone. Reen was treated as an equal for the first time in her journey. It was a revelation, 'a glimpse of the possible.'

Just a month later, with the amoebas quiet, Reen was scheduled for a corneal graft. It was an attempt to restore sight to her eye. And it worked. For two glorious weeks she could see. She would read the names of shops to her husband as they walked near their home, of number plates as they drove. A little over two years into her ordeal, full vision was back. No more carefully placing herself on the right side of friends while walking, no more fear of falling on escalators of the London Underground, no more drugs to administer throughout the night, and above all, no more excruciating pain.

But then the fog returned. The amoebas had survived the graft. It was back to square one, an all-out offensive to try to kill them off. Attacked immediately, the extent of the infection did not return, but her sight was gone. I cannot even begin to imagine how she must have felt when the darkness returned. That might have finished me.

But not Reen. In October 2013, Moorfields Eye Hospital hosted a Cornea Patient Day, and she was there. Most of the event was taken up by presentations on advances in the science. But at the end of the day, the attendees split into groups by their condition, and were given the opportunity to talk among themselves. Reen, despite her pain and distress, took the pen and pulled over a flip-chart board she spotted. By the end of the session, it was populated with the questions those in the room would have loved to have had answered from day one, information they desperately wanted but could not find anywhere. They organised themselves, allocated tasks, and then worked with their clinicians to produce co-written patient information. They were no longer merely patients to be investigated, interrogated, even blamed for their condition, and Reen had become an organiser, shaping the community.

A year on, the patient information was out and pushing Dr Google's scariest down the rankings; that Facebook group had gone from a place of fear to a constructive and well-informed community of support; and the International Standards Organisation was even beginning the long process of introducing 'No Water' warnings on all contact lens packaging. In 2015, Reen would win a prestigious Sheila McKechnie Award as Health and Social Care Campaigner of the Year and be made a Health Service Journal Patient Leader, one of 50 patients leading change in the National Health Service; in 2016 she would add Campaign of the Year from the Royal National Institute of Blind People.

When Reen and I got back in touch, I was back at university at age 31, midway through the first term of a Masters degree in ethical philosophy. My ideas about consumerism and citizenship had begun to take shape, and I'd decided these studies would help me figure out what philosophical insights underpinned them. When we found each other again, five years after we had last seen each other, nearly three after her falling ill, Reen still had her pack of drugs and drops. She was still in constant pain. She had taken a battering, and there was more to come: a second cornea graft would follow the next year, this time removing the amoebas and meaning the constant medication could stop at last, but not giving her back the sight in her eye. But through all the pain, anxiety, and trauma that would never be wished on anyone, my friend had also grown. She had found something very special inside herself, and she saw in my ideas a framework that might make sense of her experience, and unlock what she had found for many more. Soon she had hauled me out of academia, and we began the work of the New Citizenship Project together, basing much of it on her hard-won life experience.

BILLY
Grimsby, England

'Our mams would've gone,' said his neighbour Tracey to Billy Dasein one afternoon in 2017. Until she pulled that card on him, Billy wasn't intending to go to the meeting, which had been called by a local councillor. In his fifties, with rings and studs in both ears and a further stud in his right eyebrow, Billy isn't much for official efforts, The Authorities.

With the councillor's meeting Billy anticipated just another in the steady stream of opportunities for a collective wallow in the problems of the East Marsh, a neighbourhood ranked the 25th most deprived of 33,000 across England,[35] itself part of the town of Grimsby that had recently been voted the worst place to live in the country.[36] That summer, Rutland Street in the East Marsh hit local headlines on multiple occasions as the epicentre of Grimsby's misery: drug dealing, antisocial behaviour and vandalism had escalated to the point where many residents were afraid even to leave their homes.

Grimsby charts a proud history back over a thousand years, and as recently as the 1950s, even the 1970s, it was buzzing. In those days it was among the world's busiest ports, and boasted the world's largest commercial fishing fleet. Situated on the east coast of England, on the south side of the mouth of the vast Humber River that divides the counties of Lincolnshire and Yorkshire, the trawlers would head out to the north, to the rich fishing grounds off the coast of Iceland. They would bring back enough cod and haddock not just to feed the English – and establish fish and chips as the national dish – but for export to much of western Europe. And it wasn't all work. Within a few miles, past the old industrial dockyards through the residential streets of the East Marsh (its name bestowed by marshland which was drained so that houses could be built for the dockworkers), lies the long sand beach of Cleethorpes, one of Britain's original 19th Century seaside resorts.

As Britain lost its imperial swagger, though, Iceland was one among many nations to become more assertive in defence of its own interests. Over three decades and three conflicts – known as the Cod Wars – that saw the deployment of both navies and several times came perilously close to full-scale military engagement, Iceland and then the European Union gained control over the seas, and Grimsby's fishermen found their catch drastically reduced.

The town still has several big employers, in fish and other food processing, and in the chemicals factories that sprung up along the Humber in the 1960s and 70s, but the jobs are of neither the kind nor the quantity to sustain the pride of the descendants of generations of fishermen. In 1970, there were 400 trawlers working out of Grimsby docks; in 2013, there were five.[37] It should be no surprise, then, that Grimsby was an important stop on Boris Johnson's 'Get Brexit Done' tour of key constituencies in the last days before the 2019 General Election: Grimsby is the perfect example of the 'Red Wall' Labour-voting seats that made the historic switch to Conservative representation, and the perfect symbol of the fault of outsiders for the country's malaise.

Grimsby's decline provided a background to Billy's life. 'I didn't go to school most of the time, just skipped after register was taken. When I look back now,' he says, 'I wonder what would have happened if there weren't other education available later. Well, I don't wonder actually. I know. I'd be in same place as most people round here are now.' He speaks with the unique Grimsby accent, clipped, almost Yorkshire but not quite, definite articles definitely optional.

Finding little to relate to at school, he took an apprenticeship as a pipe fitter at the age of 16, to the relief of his parents. 'Getting a job was the single most important thing a person could do in that culture.' He did that work for a decade, the 'dreary flatness of unchallenging employment,' until returning to school as a mature student. After the deeply hierarchical environment of the factory, school brought Billy to life. In the 1980s, Humber Lodge in Grimsby offered an extensive range of courses, including a two-year diploma in higher education, designed primarily (in the gendered way of the time) for women looking to train or retrain as teachers after having had children. After earning his own degree, Billy became a teacher there until the closure of the Lodge in the late 1990s, which happened as the slow encroachment of the commercialisation of education made its way to Grimsby.

Billy went adventuring abroad for a few years, setting up a school in Oman and teaching English in Poland, until a job at Birmingham University offered him the opportunity to start the doctorate that he'd been pondering. Only in 2013, upon the death of his mother, did Billy return to Grimsby, to look after his father amid the onset of dementia. It was one of those catastrophes that evolves into a blessing, as Billy finally faced up to the challenges in East Marsh, beginning with the 2017 meeting he intended to skip.

With his mother's departed spirit invoked to prod him along, Billy headed to the meeting convened by a local councillor, Steve Beasant. It started out pretty much as Billy had expected: 'A load of people complaining, throwing stones at people who were doing their best but couldn't fix it.'

Billy had been reading about the Broken Windows Theory, originally proposed by conservative political scientists James Q Wilson and George L Kelling in an article in *The Atlantic* in 1982, and popularised by Malcolm Gladwell in his book *The Tipping Point*. Hugely influential, one of the most cited articles in criminology, the article opened with the sentence 'A piece of property is abandoned, weeds grow up, a window is smashed.' [38]Setting out a logical-sounding progression in which smaller crimes beget larger crimes, it would become the basis for 'zero tolerance' policies that focused on people committing petty crimes like graffiti and loitering; it led to extensive 'stopping and frisking' – and often arrests and convictions – that overwhelmingly targeted boys and men of colour. The Broken Windows Theory – like the 'superpredator theory' that came a decade later from a neoconservative Princeton professor named John Dilulio, but embraced and popularised by the very same James Q Wilson – has since been fully debunked as yet another justification for white supremacy as expressed in unjust policing.[39]

However, a baby might have been disposed of with the broken windows bathwater. The abandoned property and the rampant weeds described in the first part of the opening sentence were lost when all the focus went to the criminal act of smashing the window (everyone loves to focus on the criminal). Yet according to the research,[40] general disrepair and blight are indeed factors that correlate with the suffering (and yes, crime) of a community. Of course, they're factors that are much more challenging to address, often reflecting decades of underinvestment from both the public and private sector, often in places like Grimsby where industry once thrived but fled due to a combination of factors: international trade agreements, and cost-saving measures on the part of corporations relocating to places with less protected, less dignified labour. The authorities can't so easily arrest that and stick it behind bars.

East Marsh had certainly reached the point of looking – and indeed smelling – uncared for. I first came across Billy's story in an article titled *A Dead Dog in the Bathtub*. In an abandoned house on Rutland Street, the said dog lay in a bathtub for months. '[It] only became obvious when bluebottle flies started coming through into next door.... Police had gone in to make sure there was not a human body inside, then left again, boarding up the door behind them and leaving the dog for someone else.'[41]

Billy stood up at that meeting and called on the better natures of those present. I'm fairly certain his speech was more inspiring than he recounts it to me, three years later. As he tells it though, 'I just said we should get together and clear up, simple as that really. I got a round of applause, so I said to leave me your name if you're up for it. I got 30 names, 16 turned up for the first clear up, and the rest is history really.'

'The rest' was the founding and subsequent growth and flourishing of East Marsh United. What began as a one-off clear up quickly became a regular affair, meeting fortnightly to sort two or three streets together (including getting rid of that dog, no thanks to either the landlord who owned the property, or to the police); then things began to develop further. By the end of 2017, the growing 'EMU' team had registered the organisation as a Community Benefit Society, and the first issue of The Proud East Marshian, a monthly community magazine, had gone to print. Planning soon began for a Sun and Moon Community Arts Festival. The clear up squads evolved into fortnightly open meetings (though the clear ups still happen), with rotating chairmanship, and grants started to come in, allowing for Billy to take a small salary and then another paid organiser to be recruited.

As the work of the group continued, though, a root cause of the problems became apparent: housing, and in particular the ownership of housing. With Britain's sustained property boom, housing stock has become a major investment industry, with landlords proliferating over decades in tandem with house price rises. A good number of those landlords live elsewhere in the world, investing their wealth in premium London apartments and townhouses and frequently leaving them empty, furthering the housing crisis. But Grimsby and the East Marsh exemplify the relatively untold story of the damage at the other end.

'You can still buy a house in the East Marsh for ten grand or so,' Billy tells me, 'and you don't need to make much in rent to cover that. So people do it, and trust me, they don't give a shit about the state of the property. When they can get tenants, they can't get anything fixed. When they can't, they leave them to rot.'

It's clear how this sort of situation leads to the environment Billy described, and the resonance of the Broken Windows approach. But he and the EMU team are working on a solution to this, alongside everything else and in line with their 'OUR problems, OUR solutions' philosophy. EMU is becoming a social housing provider. Pooling resources, it's already raised the money to buy several houses in the East Marsh, and let them affordably and responsibly to new tenants while developing a revenue stream for the organisation at the same time.

When I visit and walk the streets with Billy on a rainy day in June 2021, he shows me with pride the street the EMU team thinks could be a model for the area. Wide pavements, small but enclosed front gardens where bins can be kept, a few trees. Nothing crazy, and entirely possible. The residents of several other streets are talking to EMU, and indeed joining the group, as they seek to pool funding and support one another to make it happen. Then he takes me inside EMU's latest purchase, just around the corner from his own home. This he does to show me the scale of the challenge still outstanding. With no front gardens and narrow pavements, bins are out on the street, several of them spilling. The outside of the house has been painted, but behind a locked outer screen, it's immediately clear what it must have looked like. The old front door is still on, a four letter word beginning with C blazoned across it in blue graffiti.

I come away chastened but deeply hopeful. Billy and the EMU team have a task on their hands. But they're up to it, and more importantly, up for it.

CITIZENS EVERYWHERE

The headlines of our time are all about decline, decay, catastrophe, tragedy. And they are true. But there is another truth. Dig a little, get beneath the surface, and we find people like these across the world, in every neighbourhood and every sector of society. This is why their stories, though uniquely individual, are at the same time remarkably familiar; we all know someone like this, or have heard similar stories. They didn't receive special training. None of them arrived on the planet with exceptional wealth or connections, at least relative to their compatriots. Most of us would recognise ourselves in at least one of their stories, in at least one moment.

Each might seem to embody a 'type,' a common label: Activist, perhaps, for Immy, or Aid Worker; Hacker-Entrepreneur for Bianca; Community Leader for Kennedy; Victim-Survivor for Reen; and Anarchist, perhaps, for Billy. I'd like to reframe these five people as practitioners of the Citizen Story: my emblematic Citizens. They are the force that makes the transformation of our world possible, seeding the new institutions and processes that could carry us into the future even as the old story collapses around them. As we start to see with Citizen eyes, we begin to perceive a world filled with possibility. Citizens of every race, colour and creed are finding one another, getting organised, building tools, and building the future. Immy, Bianca, Kennedy, Reen and Billy show the range of this phenomenon, each representing one of the millions of green shoots forcing their way through the cracks in the concrete, each defying the headline narrative of decline that dominates our time. The stories I could have told are enough to fill several books in their own right; all I can do here is give a sense of what lies behind these five.

Immy's work in Birmingham draws our attention to city-scale. Here, a vast array of institutions are at least to some degree within reach. Citizens are increasingly questioning, reforming, or replacing them. In place of Immy and Birmingham, I could have introduced Robert Bjarnason and Gunnar Grimmson, the game designers who returned to their home city in the aftermath of the Icelandic financial crisis and built a participatory democracy platform called Better Reykjavik that has brought hundreds of Citizen ideas into the operation of the city and allocated millions of euros according to Citizen-defined priorities.[42] I could have travelled to Liege in Belgium, where a whole swathe of land around the city has been transformed in the hands of an ever-growing movement of Citizen-owned food cooperatives; what began impressively enough with one community vineyard raising two million euros of local capital has blossomed into no fewer than 14 distinct organisations, with more on the way.[43] I could have introduced you to the Citizens working to renew and reinvent Mexico City or New York City, Medellin or Manchester, Bangalore or Barcelona, to name just a few.

Bianca epitomises a breed of Citizen-crusaders who refuse to accept that nationalism and white supremacy are the inevitable rising tide of nation-state politics. Analogous work is going on in almost every country in the world. In New Zealand, for instance, Wellington's Enspiral collective came together during Occupy Wellington in the heady days of 2011. Its members have since committed themselves to building the tools that could enable nation-states to evolve rather than demanding their collapse – such as Loomio for collaborative decision-making, and Cobudget for making collective budget allocations. Both tools allow for members located anywhere to participate effectively. The Enspiral collective itself has spread all over the world.[44]

In Kibera, Kennedy shows us the capacity of Citizens to start from less than nothing, and even he is not so exceptional in his achievements. The vast Nakivale Refugee Settlement in Uganda, for example, is home to more than 100,000 displaced people. While refugees await resettlement, often for years and with little to occupy their time, their human potential can go to waste. Here the designer and artist Patrick Muvunga, who himself fled persecution in his native Congo, upcycles waste to build shelters out of plastic bottles, jerrycans, and UN-supplied tarpaulins. Patrick also founded Opportunigee, an incubator for social entrepreneurship 'by refugees, for refugees.'[45]

In Rojava, the autonomous zone in northern Syria, the Women's Defence Units (YPJ) are behind a multi-ethnic egalitarian experiment with decentralised self-governance, respect for minority rights, and plans for ecological sustainability. They have banned child marriage, forced marriage, dowry and polygamy, and criminalised honour killings, violence and discrimination against women. 'In women, the fire for freedom burns,' says Zeynep Efrîn, a YPJ commander.[46]

In the favelas of Rio de Janeiro, Rene Silva is helping to fill the vacuum left by the authorities. He began reporting the events of his favela when he was a young teenager. Now in his mid-twenties, his online news site Voz da Comunidade employs

16 young journalists who receive funding from local NGOs but maintain all editorial control.[47] During the pandemic the site has disseminated critical information and coordinated the distribution of life-saving resources.[48] For all the distressed hand-wringing from wealthier nations over the state of slums, refugee camps, and indeed immigration, Citizens in these situations are proving that their proximity to local challenges makes them by far the best experts for the tasks at hand – whether they hold the legal status of citizenship or not.

Reen's story opens the door out of the house of anything we might think of as politics, and into the big wide world where we can see Citizens working away at the edges to change pretty much every aspect of society. In any other sector or industry or even organisation, be it a public institution, a charity or NGO, or even a business, I could have found innumerable 'intrapreneurs', employees acting as Citizens of their own organisations, working to reshape and open them up from within. At the venerable Bank of England, for example, there is the little knot of Citizen-employees responsible for creating *Bank Underground*, the open blog where staff share views that often challenge prevailing policy orthodoxies under such titles as *Less is more: what does mindfulness mean for economics?*[49] The same group has initiated a series of regional citizens' assemblies; these operate as an upstream input into the Bank's policy development process, based on a recognition that simply looking at the numbers is no longer enough, and that the expertise of lived experience – the insight of Citizens – contributes at least as much, and sometimes more.

Finally, there is Billy. From the outside he and I look like the adversaries in the fight: myself a millennial (just about), a member of the south England metropolitan liberal elite (most certainly), Oxbridge educated, with every privilege in the book; him a working class northerner, a child of the 1970s, somewhere between socialist and anarcho-communist in political leaning, born in a town whose fortunes were undeniably undermined by the circumstances of Britain's entry into Europe. With only our pale skin and male gender in common, we are supposed to be adversaries, with those whose skin is of a different hue destined to be the collateral damage in our fight as they have been for centuries. But I look at Billy more like a brother than an enemy. Our fascinations and passions are the same, not opposed. In our first follow-up conversation after a chance meeting on a random conference call, I learned more from him about the work I am doing, not just about its practice but about its intellectual foundations, than in any other hour I have ever spent.

2. CITIZENS BY NATURE

When we get beneath the surface of the daily headlines, we find Citizens active across the globe, across place; similarly, it doesn't require much digging to get beneath the surface of the conventional history of our species, and find Citizen behaviour across time. The story of humanity was largely taught to me as one of Great Men doing Great Deeds, with everyone else largely irrelevant, but there is a far more nuanced picture to be painted. I'd say that much of the history of our species can be better understood as the repeated rising of our deep Citizen inclination, again and again over the centuries and millennia. Citizenship is human nature: an intrinsic inclination, always bubbling, often suppressed, never wholly conquered, and now at a moment of huge opportunity.

To see this, we first have to remember that our history starts long before writing and state-based organisation, which is where others often begin, yet which limits the scope to less than five per cent of our actual lifespan. Our history starts instead with small, mobile, dispersed bands of hunter-gatherers. These represented our primary mode of operation for tens of thousands of years, and the evidence suggests that in this mode we were both egalitarian and highly participatory in our decision-making. This is best captured by the sociologist Ronald Glassman's evocative phrase 'campfire democracy,' reflecting the manner in which members of a band gathered, often around a central fire, to make the big decisions collectively as peers.[50] There is a growing body of evidence that in pre-colonial Africa, indigenous Australia, and among Native Americans, 'government by discussion' was the norm: less the despotic tribal chieftains of lore, and more the Citizen. In one of contemporary humanity's rare remaining hunter-gatherer societies, in southern Afria, a Ju/'hoansi man was asked why his people seem to have no 'headman,' and his response was, 'Of course we have headmen. Every man is head of himself.'[51]

If the Citizen is acknowledged anywhere in conventional Western history, it is in Ancient Athens – and with good reason. Starting early in the 6th Century BCE and continuing (with a number of invasions and interruptions) over the next 300 years, enlightened Citizen leaders such as Solon, Kleisthenes and Perikles pushed power out and down through society, rather than holding tight to the reins. Concepts and processes emerged that still shape our society today. The terms *demokratia* (literally, 'people power') and *politike* ('affairs of the city-state') were coined, the concepts given substance and coherence as a result. Neighbourhood forums were created, with clear boundaries and powers. The process of sortition emerged, which saw representatives, judges, and other key posts randomly selected by lot, as opposed to either inheriting their roles or being elected to them. This still underpins the selection of criminal juries across the world, and is a core feature of the deliberative processes (such as Citizens' Assemblies) that are growing in credibility and prominence in today's Citizen emergence. Athens was flawed – these opportunities and rights only extended to free men, and therefore excluded not only the female half of the population but also slaves – but it remains the case that at the peak of Athenian democracy, as many as 60,000 Citizens were making the decisions and setting the direction of their society together.

For all its significance, however, Athens was not the site of the immaculate conception of democracy and Citizenship that my Cambridge Classics degree would have had me believe. Not only were the roots present in indigenous cultures, but participatory processes and institutions also continued to play an important role in almost all early states, from Mesopotamia to the Indus Valley. Nor did democratic darkness descend after the classical era of Greece and Rome, as is often said. Medieval Islamic empires used political systems similar to what we would now call democracies, and there were further Citizen moments in Europe too. Iceland, for example, was an uninhabited island until around the year 870. According to the Icelandic sagas, a steady stream of settlers arrived from Norway over the next 60 years, many of them fleeing the brutal rule of King Harald. Wanting an alternative way to rule themselves, they created a network of 13 district assemblies which convened every spring to settle local disputes, and a two-week long annual convening of all male citizens (*Þingmenn,* literally 'assembly men') on the open land of *Þingvellir* ('Assembly Plains') a little way east of modern-day Reykjavik. This evolved into Iceland's national parliament, and is today considered the world's oldest surviving parliament.

In the research for this book, I have particularly enjoyed re-learning the history of my own country through a Citizen lens and finding some reasons for pride. A Citizen history of Britain focuses, for example, on the signing of Magna Carta by King John in 1215. With its restrictions on the monarch's royal powers, it has been called 'the greatest constitutional document of all times – the foundation of the freedom of the individual against the arbitrary power of the despot.' It ensured protection from illegal imprisonment, access to justice, and limitations on taxation, first and foremost for free men, but also with provision for basic rights for those in bonded labour. Fast forward to October 1647, and the Putney Debates offer another important Citizen moment. With the Civil War won and Charles I under house arrest, the New Model Army came together at St Mary's Church in Putney, south west London, to debate how to run a kingdom without a king. There, Colonel Thomas Rainsborough first proposed the concept of universal suffrage, regardless of property ownership – although still only among men. The Westminster parliamentary system is another Citizen creation: for all its increasingly apparent flaws, it served the important purpose of reducing the unelected head of state to a status that was ceremonial only, and would become known as the Mother of Parliaments as a result.

Another Citizen emergence came with the so-called Golden Age Pirates, who ruled the waves from around 1690 to 1725. Following the cinematic tropes of Captain Jack Sparrow and the like, I used to think of pirates as charismatic characters but ultimately violent criminals and outlaws. Thanks to Sam Conniff's book *Be More Pirate*, I now see them as Citizen heroes.

Conniff pulls together evidence that many of the stories of plank-walking and arbitrary violence were fabricated, by a monarchy that had only just re-established itself in Britain after the Civil War. Anti-pirate propaganda was a deliberate effort to undermine the legitimacy of a renewed threat to royal power. In reality, pirate ships were more often than not places rich with Citizen innovation. There were clear limits on the powers of Captain and Quartermaster, with the two set up to countervail one another, and clear guidelines for the equitable division of loot. There was often a one-person-one-vote process for the big decisions. Pioneering social insurance practices also developed on board, with pirates injured in battle receiving a payout from the ship's common pot of money. Indeed, it's possible to trace the lineage of the Golden Age Pirates and their ideas not just back to the Civil War, but forward to the birth of the cooperative movement and even to some of the nobler and more equitable aspirations of America's Founding Fathers.

Of course, it is not just British history that can be revisited in this way. The truth is that every nation and every people have Citizen moments to honour in their past, just as there are emblematic Citizens to be found in every neighbourhood, town and city across the world today. The simple truth is that this is who we are, as humans – or at least, who we are capable of being, if the conditions we create together allow it, and if we choose to be.

CITIZENSHIP AS PRACTICE

Fundamentally, Citizens are humans who want to shape the world around us for the better, and who claim and where necessary demand the right and the means to do so. To be a Citizen is to care, to take responsibility, to acknowledge one's inherent power. To be a Citizen is to cultivate meaningful connection to a web of relationships and institutions. Citizenship benefits from a free and expansive imagination, the ability to see how things could be, not just how they currently are. To be a Citizen implies engagement, contribution, and action rather than a passive state of being or receiving. This is the most accurate sense of the word etymologically: Citizens are literally 'together people,' from the Latin, humans defined by the very fact of their togetherness. We tend to think it is the other way around, but 'city' in fact derives from the Latin *civitas*, which literally translates as the citizenry. As such, a city is literally a place where people are together. Likewise, 'civil,' 'civilised,' and 'civility' are all words for the art of relating and working and ultimately doing life together.

Being a Citizen is a way of life, almost more a verb than a noun. A practice. One leading Citizen voice has taken this logic to its full conclusion: in the spring of 2020, the American cultural critic Baratunde Thurston launched his podcast called 'How to Citizen,' explicitly reimagining 'Citizen' as a verb and providing inspiration to his listeners on how to participate in collective action and governance in its widest sense. 'This is not a show about how a bill becomes a law,' he says. Instead it's about 'who has the power to determine the quality of our lives. We believe the correct answer is all of us.'[52]

95

As a practice, our capacity for Citizenship is infinite. We do not each have a 'civic capacity,' a fixed amount of participation we have available that must be used wisely in case it runs out. Instead, we can think of Citizenship as a muscle, something we build the more we exercise it.

All of this is distinct from another interpretation of the term, what I have come to think of as citizenship-as-status. Used in this way, the language of citizenship is heavy and charged, a powerful carrier of the xenophobic opposition of 'Us' versus 'Them' that underlies nationalism and protectionism. In this framing, citizenship determines rights to work and healthcare, freedom of movement, opportunity to vote, and protection from deportation. Policies, law and culture around migration, immigration and naturalisation determine who gets these and who does not.

Citizenship-as-status is very much a noun – a legal construct, a possession a person either owns or does not. This legal construct is increasingly unevenly applied, selective and subject to manipulation, privileging people who represent economic advantages, and reflecting legacies of discrimination based on race, religion, and more. In Britain, we talk about people being in the country 'illegally.' Highly targeted social media advertising, largely invisible to those it was not intended to persuade, pitched Brexit as protection against a siege of mostly Muslim immigrants, as a reassertion of white English identity. In the USA, those who would close the borders use crass language like 'illegal aliens' or just 'illegals.'

In reducing it to a possession, citizenship-as-status taints the essence of Citizenship. This in turn creates the space for abuse of fundamental human rights and dignities, dehumanising those who do not have it. The fast-growing number of stateless people worldwide who have been denied or stripped of their membership in a nation are not official citizens. People who are permitted to cross borders into wealthier regions to perform undesirable or unprotected kinds of work – like migrant farmworkers and construction workers, or caregivers, or sex workers – are tolerated (exploited) but extended none of the protections of citizenship-as-status.

At the other end of the spectrum, citizenship-as-status is a possession that can increasingly be bought: more and more people hold dual or triple nationalities, sometimes because of ancestral heritage or true connection or allegiance (as by long-time residence or marriage) to the nation in question, but increasingly based on raw wealth. The phenomenon of so-called 'golden passports' seems to be expanding. In April 2021, for instance, investigatory journalists found that a significant number of rich individuals received Maltese citizenship after spending just days or even hours on the island.[53] The attraction, of course, is that Malta is a member of the European Union; Maltese citizenship is European citizenship.

These dynamics make citizenship-as-status less relevant, but they make the true Citizen more vital than ever. The language of Citizenship is crucial, and the contest for its meaning is one that must be engaged in, not conceded. The Citizenship at the heart of this book is not a question of what passport we hold, it is a story of who we are as human beings: a question of what we can do, and what we should. In these terms, there is no human – regardless of their papers, passport, or criminal record – who cannot be a Citizen, and no limit to what Citizens can do.

WAY BEYOND VOTING

The five Citizens presented in the previous chapter embody one especially important point: the tasks that Citizens can and do take on go well beyond the one activity generally associated with citizenship-as-status, namely, voting. In that view, our civic role and responsibility is limited to the single act of choosing representatives. Our contribution is to enter the hallowed booth once every few years in order to cast our vote in the General Election, the occasional referendum, and at most in local council elections. Beyond that, a tiny few might get involved making phone calls or canvassing door-to-door to support a candidate, or sign people up to vote and remind them on the big day. Some might respond to a consultation, sign a petition, or pen a complaint. But that's about the extent of it. There is nothing wrong with these actions in themselves, indeed I believe passionately that elected representatives have a vital role to play. But the current situation, where this is the beginning and end of civic engagement, is deeply problematic, for a number of reasons.

The first problem returns directly to the exclusionary nature of citizenship-as-status: a number of people who reside in a given nation are not permitted to vote there, despite being equally if not more significantly affected by the results of elections. There are plenty more who doubt, often with good reason, that voting makes much difference anyway. Uninspired, frustrated, disappointed, a good number of people neglect – or adamantly refuse – to vote, even in general or presidential elections. Under existing electoral systems in the US and Britain in particular, plenty of people who live in 'safe' seats or states don't vote on the basis that doing so will make no difference. In most places where voting is not compulsory, some 30-40% or even more of us don't vote even in major elections, and when the elections are local, as many as 70-80% of us sit them out. Then – often accentuating the effects of these problems – there is the potential for corruption, abuse, and manipulation during the disproportionately focal moments that election campaigns represent: something that has always been present, but has in recent years escalated beyond all previous comprehension.

These are challenges that can and should be addressed by electoral reform: expansion of the franchise; introduction of proportional representation and alternative voting systems; oversight and perhaps breakup of the major social media platforms (*something we will return to in Chapter 6*). These things are essential to the health of our democracies. But these challenges are just surface symptoms, and these prescriptions are not the primary focus of this book. We need to keep peeling back the layers to get to the root of the problem; we are not done yet.

The next issue with voting lies in how we are invited to make our choices: the question to which we are invited to see our vote as the answer. Increasingly, that question, more or less explicitly, is: 'What is in your individual self-interest? What in your view is best for you?' This has become so taken for granted that my drawing attention to it might seem unnecessary. But the fact is that this very framing of the choice, before any of the money or dirty tricks begin, is a corruption of what voting is all about. In order to contribute constructively to the governance of a nation, voting must operate on the basis of what has come to be known as the 'wisdom of the crowd.' For each of us to contribute to the pool of collective wisdom, it is critical that the question we instead answer is: 'What is in our collective interest? What is best for the nation as a whole?' When we aggregate our multiple perspectives in answer to this question, we stand a good chance of getting to a better answer than any of us would alone. When instead the question becomes one of self-interest rather than collective interest, things inevitably fall apart – because we are no longer contributing our different perspectives to the same question. Instead, each of us is answering our own question, with the system attempting to aggregate that. This incentivises political parties, contesting for our votes, to segment and divide us, calculating their efforts to win just enough to tip the majority – no matter how the rest of the population might feel. As this dynamic deepens its hold over time, competition, division, and aggression become increasingly inevitable – as we can see every-where when we look around us.

There is one more layer yet to the problematic nature of a sole focus on voting and electoral politics as the expression of citizen-ship: the very idea that formal government is what it's all about. This is politics as a spectator sport at best, rather than a partic-ipatory pursuit, and as such is the specialised realm of experts who are a breed apart from the rest of us, just like professional athletes. Elected officials, civil servants, lawmakers, leaders of governments and architects of world trade agreements tend to have attended one of a handful of select universities and even schools, to hold one of three or four degrees, 'turning profession-al' as early as their teenage years. As a result, too many tend not to know the lived experience of the majority of people on behalf of whom they are tasked with making decisions. It is a whole other world, with no place, really, for the rest of us: all we are good enough to do is choose between options that someone else has decided to offer us.

This is where the deepest damage is done. Yet what truly matters most is what the rest of us do every day, the convivial and pragmatic dynamics of making choices, compromises and agreements in every sphere of life. This is the domain of Citizens, and it is a space in which voting must take its place as one action among many, not the sole and defining contribution.

Official citizens only do capital-P Politics once every few years; Citizens do small-p politics pretty much every day. As Citizens, we are not our votes: we are our ideas, our energy, our resources. To convey that voting represents the only or even the most meaningful opportunity for contribution shuts almost all of us out – to a greater or lesser extent – and then shuts the remaining few in, piling on pressure that they cannot possibly sustain. For all of us and in every sense of the word, this situation is unsustainable.

LESS THE NATION, MORE THE HOME

Citizenship-as-status is a possession some have, but many lack; Citizenship-as-practice is something we can all do. Just as official citizens have only their votes to contribute, they have only their nationality with which to identify and to defend. By contrast, true Citizens have homes: communities of place and interest, often overlapping, in which they engage, and to which they contribute in myriad ways.

As Citizens we begin our journey by establishing what home is: the domain (or domains, since many have the bandwidth for more than one) where we belong. Home can be local, a matter of the city, the neighbourhood, even the street; or we can change the angle and see 'the place we belong to' as a community defined less by physical space and more by passion, interest or shared experience; or we can zoom out and see it as encompassing the whole world and even beyond. This is the idea of world-mindedness that is so essential to Citizen action on the climate emergency and so much else, based on the recognition of our fundamental interconnection and interdependence.

In my own life, I feel a clear and strong sense of home in the place where I live, in my neighbourhood and town, as well as an allegiance to country (which is to say, an allegiance to a vision of my country's highest potential). I'm one of those people who hasn't strayed terribly far from the place where I was born, or indeed from where my parents and grandparents were born, raised their children, and settled down. I have roots here, in a fairly literal sense. There are many people, however, for whom things aren't quite that simple.

Between mass migration and mobility, home can be an elusive concept. Globally, the number of people who left their home countries to make a life elsewhere increased from 150 million in 2000 to 272 million in 2019.[54] It is projected to rise to 405 million by 2050.[55] Some 41 million of them have been displaced due to conflict or violence.[56] An ever increasing number leave their homes due to inhospitable climate or extreme weather events. Most – nearly two-thirds – left their places of birth to find work in larger, stronger economies. More than five million Britons also make their home elsewhere.[57] The term for them is generally 'expats' (expatriates) – language reserved for the mobile person from an affluent country – and more recently, among freelancers of Generations Y and Z, 'nomads' or 'digital nomads' – as opposed to 'migrants' or 'refugees.'

Children who spend the formative years up to the age of 18 away from their parents' place of origin blend the cultures of their parents and place of birth with the places they live. Some are children of executives of multinational corporations; some have parents in the diplomatic corps or the military; others are children of migrants or refugees, or children adopted by parents from another country. Third Culture Kids (a term coined in the 1950s by US sociologist Ruth Hill Useem) dread the question 'where are you from?' and even worse, after their simplified answer, the follow-up: 'But where are you really from?' 'It's complicated,' they tend to answer.

Given the dynamics of mobility today, most Citizens must first grapple with the question of belonging and connection – finding and creating a sense of being at home somewhere. Home, writes Stephen Jenkinson, 'means more than having memories associated with a given place. It means learning again how you and those you love and admire – in every physical, metabolic, chemical, mythical and spiritual sense it can be meant – are made of the things that make the place you belong to. That is the alchemy of belonging.'[58] Being at home is not a feeling; being at home is a skill,[59] maintains Jenkinson, who came to these conclusions after many years supporting people at the brink of death, having learned much from indigenous peoples about rootedness and a connection to ancestors and land.

His definition echoes the findings of Brene Brown, the famed 'shame and vulnerability researcher' known for her Texan twang and her self-deprecating TED Talks. 'True belonging,' Brown writes in her book *Braving the Wilderness*, 'is not passive. It's not the belonging that comes with just joining a group. It's not fitting in or pretending or selling out because it's safer. It's a practice that requires us to be vulnerable, get uncomfortable, and learn how to be present with people without sacrificing who we are.'[60]

One of the ways that we know a place or a domain qualifies as 'home' is that it shapes us as much as we shape it. Any place we consider a home, in turn remakes who we are, in reciprocal and cyclical fashion. These dynamics flesh out the idea of Citizenship as active rather than passive. Being at home – being a Citizen – is neither about an objective physical location we just happen to be in or 'come from,' nor even a subjective feeling; being at home somewhere involves engaging, paying attention, and relating. It is something that can and must be learned and worked at, ongoing, in order to be and remain authentic and true.

Citizens can only succeed where our voices and opinions can be heard. Depending on factors like skin colour, gender, faith, ability, income and level of education, some of us may be unwelcome or even forbidden from expressing our views and taking action. Financial poverty and time poverty, which often go hand in hand, often form an insurmountable barrier to participation. We will have to address and remove all structural and cultural obstacles so that everyone can contribute, understanding people no longer as victims but as holders of vital perspectives. In the meantime, especially at the outset of our engagement, some of us may need to start with bubbles of safety before graduating to more extensive domains, or prepare ourselves for rejection and dismissal. But when we find the convergence of where we belong and where we are encouraged or at least allowed to make a contribution, the magic happens.

CITIZENS EQUIPPED

Today we are better outfitted for the practice of Citizenship than in any previous period of history, because we have the internet.

The promise of the digital age was visible very early on. In its chaotic, open, and intensely creative origins, the internet offered individual empowerment and access, cross-pollination, almost limitless connection among and between Citizens everywhere. As a many-to-many medium, it asks more of us than television or radio or the printing press, equipping us to be active in the world, capable of representing our beliefs and values through our actions, not resigning ourselves to be acted on. As the Canadian philosopher Marshall Mcluhan articulated with the aphorism 'the medium is the message,' the medium of communication itself – books, radio, television, the internet – affects who we are and how we interact in a way that goes way beyond the content communicated.[61] The media we use shape not just what we communicate but how we communicate, and therefore how we think and function, individually and collectively, right up to the scale of a whole society. When a new medium arrives, we should think of it as an 'extension' of human capacity, and as such something that affects, well, everything. Since it is a medium which in principle allows everyone to create, not just consume, we could say that the internet asks us – enables us – to be Citizens.

In the early days, the internet seemed set to deliver on that promise all by itself. This was still the case in the early 2000s, with the emergence of 'Web 2.0,' characterised by one media theorist as a 'move from personal websites to blogs and blog site aggregation, from publishing to participation, from web content as the outcome of large up-front investment to an ongoing and interactive process, and from content management systems to links based on tagging (folksonomy).'[62] First we had Wordpress, then we had Napster, then we had Meetup; we could create and share words and music and find community. The tools were being built, and the new society would come.

To some extent, the promise was fulfilled. Digital technology has opened up an astonishing number of new modes of participation, as well as reinvigorating many of the old – open innovation challenge prizes, volunteering programmes, participatory budgeting, crowdfunding, to name but a few. As we see with particular clarity in Bianca's story, but true across the world, political entrepreneurs are harnessing the deliberative, creative potential of this medium to experiment with many-to-many, participatory models of politics: the Open Ministry in Finland, Better Reykjavik in Iceland, secondgov in the United States, and Loomio from New Zealand are all excellent examples of this. They're all about enabling deliberation and co-creation of policy. They're harnessing the potential of the internet to bring politics into the Citizen Story.

But on a broader view, something has gone very wrong. The internet has become increasingly abused as a tool to undermine Citizenship, not unleash it. One grave disappointment came with what was once called the 'Sharing Economy,' now better known as the 'Gig Economy.' At the outset, it presented not just an economic hack (getting more utility and value out of under-used resources), but a social and environmental win, empowering the underemployed, connecting strangers and reducing consumption. But soon the gravitational pull of business-as-usual pulled the new models in and down. Airbnb, Uber and the like started out as beautiful dreams. But rather than transforming transactions into human relationships, they've done the opposite. Mediated by these platforms, human interactions come to feel more like transactions. Ultimately, they have reduced us to mere Consumers again, in this case not just of online content or even the products of corporations, but of each other.

Nowhere is the perversion of the internet's Citizen potential clearer than in the case of Cambridge Analytica. The data analytics company successfully manipulated undecided voters, particularly in contested regions, during a number of political campaigns, culminating in LeaveEU (Brexit) and Trump. As the investigative journalist Carole Cadwalladr explains, Cambridge Analytica 'profiled people politically in order to understand their individual fears, to better target them with Facebook ads, and it did this by illicitly harvesting the profiles of 87 million people from Facebook.' The earth-shattering implication of the company's 'Great Hack' is, in Cadwalladr's words: 'whether it's actually possible to have a free and fair election ever again.'[63]

We give our data away for free in return for treats, and restrict the participatory potential of this moment to choosing the colour of our trainers. Meanwhile, nefarious actors use that data to shape our societies.

Despite its disappointments and dangers, though, the internet remains a key reason why all this Citizen activity is happening now, to an extent never seen before, and a key reason why the opportunity we have now is greater and more global than at any previous moment of Citizen emergence in our history. It still has the potential to drive and sustain a many-to-many society, where we can all shape and produce society itself, where we can all be Citizens. It is not, however, going to do it for us automatically. We will have to make sure we have a truly Citizen Internet: everyone must have affordable access, free from the control of corporations or governments, and the right to both privacy and security. Even with that in place, the Citizenship we need won't exclusively be happening online.

Yes, we have the internet. Now we need to bring ourselves.

THE POWER OF AMATEURS

As Citizens, we bring our full selves, comprised of the sum of our experiences, emotions, and expertise. We deliberately cultivate a set of skills that have been under-appreciated for far too long – but which are inherent in every one of us. These are the 'soft skills' – like listening, communication, facilitation, empathy and other interpersonal and social skills – that show up everywhere in the practice of Citizens. Whatever backgrounds we have, we repurpose them for the task.

The American marine biologist Dr Ayana Elizabeth Johnson, also a strategist for federal environmental policy, comments on the failures of the environmental movement, describing the 'enormous miss' of 'asking everyone to do the same thing: everyone march, everyone spread the word, everyone reduce your own carbon footprint,' which resulted in individual 'superpowers' being left untapped. 'Instead of all following the same exact checklist,' she says 'I encourage people to figure out the special things they can contribute.... One way to do that is to draw a climate action Venn diagram with three circles. One circle is "What are you good at?," the next is "What part of the climate problem do you want to help solve?," and the third is "What brings you joy?" And then figure out how you can work at the epicentre of that Venn diagram for as many minutes of your life as you can.'[64] Her excellent suggestion can be appropriated and extended beyond climate, to every arena: how about Citizen action Venn diagrams?

This, ultimately, is the most important lesson Immy, Bianca, Kennedy, Reen, and Billy have to offer. All lack expertise in the arena(s) in which they have chosen to engage – at least, traditional expertise. A degree in city planning (Immy or Billy), or political science (Bianca or Kennedy), or medicine (Reen). They do not disrespect or dismiss these things, but they don't see their lack as a reason not to contribute, to leave the field for someone more qualified. As creatures of action who can and want to shape the world around them, who engage daily in small-p politics, Citizens do not try to become something they are not. Instead, they bring everything they have to take action where they feel called to do so, and pull on other wisdom and expertise to complement that.

This is starkly different from the world as it is, which relies heavily on qualifications, credentials, and professionalisation. Leave things to the experts, is the message we receive. Politicians overwhelmingly hold the same degrees from the same universities; so do CEOs and venture capitalists, those who decide which enterprises to fund. Even the charities and NGOs whose work it is to address so many of the world's complex challenges, from poverty to violence to climate chaos – the 'third sector' – rely hugely on funding granted by a small group of decision-makers whose qualification is having a lot of money (which frequently isn't even earned, it's inherited). These markers are what has qualified as 'expertise.'

And look where it's got us. If this is how technocrats respond to climate emergency, it's time to bring in the actors and startup coaches, as far as I'm concerned. If the official government of Kenya is fine with people dying of preventable causes on the outskirts of its capital city, then I welcome the metaphorical mayors. If the medical profession is failing to communicate the dangers of contact lenses and people are losing their sight as a result, then maybe the advertising account managers are exactly what's needed. No, to speak to everyone's favourite examples of the need for licenses, I don't cherish the idea of driving across a bridge without a structural engineer involved, or the prospect of surgery under the hands of anyone but a medical professional, but they are not the only people I want in on the processes of designing the infrastructure of my nation and the provision of healthcare.

In his book about the 'Maker Movement,' the explosion of hobbyists and hackers that happened in tandem with the increased accessibility of technology like 3D printers and the rise of open source designs, Dale Dougherty writes:

> *I believe in the power of the amateur to do something professionals won't or can't do, or plainly, just haven't done. I believe the novice can see things that the experts miss, and do things that the business-minded don't properly value. I also believe that ordinary people can develop uncommon insights and act on them because they have not been taught the 'right' way to see things. It is a crazy and idealistic but ultimately democratic idea to believe that we can all contribute, and that art and innovation come from unexpected people and places.*[65]

We call people who don't possess the right credentials 'amateurs,' but the root of that word is from the Latin for love, and that's its own kind of qualification. Arguably, that deep care and passion – which the Citizens I see everywhere have in spades – is the vital ingredient missing among those who have been left in charge.

THIS IS WHO WE ARE

When we open our eyes to the number of people who are practising Citizenship in small and large ways, to the times and places throughout history when this instinct came to the fore, and to the tools and opportunities we have available to us today, we start to believe another story of ourselves is possible. At that point, as we go back out into the world, we crash headlong into the prevailing beliefs about human nature: namely, that each of us is only 'out for number one'; that humans are inherently greedy and selfish, calculating and corrupt. We are made to feel naive, looked upon as idealists and pointless dreamers. We are stopped in our tracks.

I have taken to starting my response to these challenges by sharing what happened in my neighbourhood during the first pandemic lockdown in the spring of 2020. One day a postcard dropped through my door. It read 'Hello! If you are self-isolating, I can help.' It gave the name, address and mobile number of someone who lived around the corner. A second identical card dropped through later that day, and another the next. In the bottom corner was the hashtag #ViralKindness. I went online and found the story of the designer who had made the cards for herself to deliver.[66]

In the article, I found out that a network of mutual aid groups had started springing up all over the country, and indeed the world. I found my local group among them, and signed up myself. Within days this group had proliferated in numbers, and self-organised into hyper-local subgroups. At the same time, a neighbour from our street sent a WhatsApp message, forming a street support group as well. I watched as food, chores, and jigsaw puzzles were exchanged. When a few weeks later my partner and I crashed while out on our bikes taking our daily exercise, ambulances took us to hospital. The local mutual aid group arranged for the collection of the bikes from the side of the road, and the neighbours, organising via the street group, took turns bringing meals for the next week as we recovered.

Was this unique to my local community? No. It was happening everywhere, across every demographic. Food banks, errand running, facemask making. Between February and May 2020, the proportion of the population who agreed that 'Britain is a place where people look out for each other' tripled, to over 60%.[67] In late March, the government launched a national scheme for volunteers to support the work of the National Health Service, setting a target of 250,000 sign ups. When 750,000 people signed up within 48 hours, the system crashed.[68] What I experienced was the rule, not the exception. And the lesson of history, according to the philosopher and writer Rebecca Solnit, was that all this was exactly what we should have expected to happen.

In *A Paradise Built in Hell*, her 2009 book about human response to disasters through history, Solnit describes how communities invariably come together during and after crises, developing new ways not just to survive but to thrive, healing old wounds, and finding joy in the process. Her book ranges from San Francisco's earthquake in 1906 to Mexico City's in 1985, from Britain in the Blitz to Vietnam under American invasion, from post-9/11 New York to post-Katrina New Orleans, and much in between, with the same core insight. 'These accounts,' she writes, 'demonstrate that the citizens any paradise would need – the people who are brave enough, resourceful enough, and generous enough – already exist.'[69] If people did not by nature want to be good, kind, and useful to one another, this is not what should happen. When disaster strikes, we should instead be reduced to our vicious nature, to saving ourselves and letting the rest be damned. Instead, we rally, we create, we collaborate, we care.

As for the widespread beliefs that selfishness, competition and status are the core inevitable drivers of evolution and therefore human behaviour? The 'evidence' of that position is being steadily undermined.

Take Richard Dawkins' 'selfish gene' theory, often interpreted simplistically as an irrefutable argument that creatures evolve and behave solely in the interest of their individual survival. Nichola Raihani, Professor of Evolution and Behaviour at University College London makes a compelling case for 'the social instinct' (the title of her 2021 book). Even if genes might be considered self-interested in the sense of 'wanting' to be passed on, at the level of the organism, cooperation is often the best strategy for survival.[70]

Or consider the idea of 'the tragedy of the commons,' that humans will inevitably drain a shared resource without top-down government restraining us. Garrett Hardin, who coined 'the tragedy of the commons' as a phrase, has been outed as a white nationalist and a racist, the motivations for his assertions significantly clearer than his evidence base.[71] By contrast, Elinor Ostrom won the Nobel Prize for Economics in 2009 after decades of on-the-ground work studying the management of common pool resources. The right conditions need to be in place, among the most essential of which is a clear set of boundaries and meaningful power at the local level. But when those conditions are present, we are more than capable of thriving together, and sustaining the resources that sustain us.

Then there are the two landmark sociology experiments that get regularly cited as proof that humans are selfish and brutish

– the 1971 Stanford Prison Experiment, where participants were randomly selected to play the role of prison guards and prisoners, and the Milgram experiment (originally in 1961), in which participants were led to believe they were administering electric shocks when instructed to do so. These findings have also been found questionable. Far from obeying 'blindly' – with the implication that humans are quite happy inflicting pain on others – participants were induced to do so, making the results nothing like 'objective' or 'scientific.' In Milgram's case, 66% of participants actually disobeyed the order to administer pain, and those who went along with it had to be persuaded that doing so served the higher purpose of advancing scientific progress.[72] 'One man said afterwards that he had persisted for his daughter, a six-year-old with cerebral palsy. He hoped the medical world would one day find a cure: "I can only say that I was – look, I'm willing to do anything that's ah, to help humanity, let's put it that way".'[73]

What the 'prevailing wisdom' of the selfish gene, the tragedy of the commons, 'red in tooth and claw' actually show us is that we are deeply susceptible to stories about ourselves. Stories are so powerful that they can be employed to redirect our innate desire to be good and cause us to take an immoral action. The work of Raihani, Ostrom and others is not simplistic; just because Citizenship is inherent in human nature does not mean we are always wonderful to each other, and everything will automatically be fine. We have the capability, but the conditions – and in particular the stories – need to be present for that capability to express itself.

When we understand ourselves as Citizens, we embrace a different story of humanity. When we recognise this, we see that we all have power, and we see tremendous potential for change. In order to unleash this, what we need to do now is understand the stories we are telling ourselves, take deep care with our language, and build from there.

PART II: SEEING OUR PRISONS

Who we are, and who we tell ourselves we are, are not the same thing.

We are Citizens by nature. Over the last eight decades, however, we have become increasingly trapped in the Consumer Story. Before that, we spent centuries, even millennia, imprisoned in the Subject Story. These stories have functioned as a kind of source code, shaping our beliefs and indeed our morality, not only guiding our behaviour but even constraining the possibilities we can imagine. Seeing the role these stories play, we can understand how the world's brutal realities and catastrophes have come to pass despite our intrinsic orientation to care and engage.

'You're here because you know something. What you know you can't explain. But you feel it. You've felt it your entire life. That there's something wrong with the world. You don't know what it is, but it's there... like a splinter in your mind, driving you mad...'

That is Morpheus talking, the mentor in the film *The Matrix*. He goes on: 'It's all around us, even in this very room. You can see it when you look out your window or when you turn on your television. You can feel it when you go to work, when you pay your taxes. The Matrix is the world that has been pulled over your eyes, to blind you from the truth.... Like everyone else, you were born into bondage, born into a prison that you cannot smell or taste or touch. A prison... for your mind.'

It's not a bad way to conceptualise 'story' in the sense I mean it: a foundational narrative that shapes every aspect of our world. Everything we build, literally and metaphorically, from the physical infrastructure of our society to our institutions and all the products of culture, have their roots in the story and reflect its logic back to us. The story surrounds us on a daily basis, providing near constant conditioning. That's why the prisons that these stories represent are so hard to escape.

In the following chapters I first describe the hallmarks of the story we're living inside today – the Consumer – and then the story that came before – the Subject. I breeze through a highly condensed history of each, allowing us to focus on its dynamics, how each story's key tenets spread from one region of the world to another, from sector to sector, until it was total and encompassing.

I am intentionally avoiding using the terms capitalism and communism. These are so laden that they overwhelm us, stultify conversation and prevent us from seeing alternatives or taking action. Nor are the mind-numbing and guilt-inducing details the point: numbers of appliances and vehicles sold, or the exponential growth in square footage of department stores or self-storage depots – during the Consumer Era – or the number of people who were subjugated and exploited during the Subject Era.

The point is to show that it was the stories underneath our society that gave rise to all of this, and then to see that we do indeed have the power to shape and change these stories.

3. WE'RE ALL CONSUMERS NOW

The most important advertisement in history was screened nationwide in the United States only once, taking over the entire first commercial break of Super Bowl XVIII. Directed by an already-famous Ridley Scott (the first *Alien* film was released in 1979), it is a truly epic film, and a great piece of storytelling. In a take on Orwell's classic dystopia, troops of brain- and colour-washed zombies parade to their seats in a great hall as a Big Brother figure drones from a massive screen. Then, through their midst, an athletic young woman sprints into the hall, carrying a giant hammer. She swings it, Olympic-style, and smashes the screen. The final voice-over rolls: 'On January 24th, Apple will release Macintosh, and you'll see why 1984 won't be like *1984*.'[74]

At one level, this was just an advertisement launching a product. The young woman represented Apple, Big Brother was IBM; the ad positioned the Macintosh as breaking open a market previously dominated by an establishment player that had little care for design, or quality, or cost. But she also represented something far bigger and deeper: the idea that, in every aspect of our lives, we deserved better than the take-it-or-leave-it outputs of establishment power. Getting what we wanted - and getting the best deal - had become a virtue. Innovation would lead to progress and well-being, and since consumer demand would drive innovation, the power of our individual choices would shape a better world. What is more, it would set us free. There would be no nightmare authoritarian society, where not just every action but every thought was limited by those in power, per Orwell. This wasn't just an ad. It was a parable: an expression of a deep, underlying story about who we are and how we should relate to one another and to society.

That underlying story is the Consumer Story.

The Consumer Story goes like this: each of us is out for ourselves, and that is the way it should be. We are individuals, narrowly defined and independent of one another; the 'self' might extend as far as our immediate family, but no further. Human nature is lazy, greedy, and selfish, but can be overcome if we set our minds to it and work hard. Our task is to earn money, spend it, and compete with one another to climb society's ladder. Along the way, we express our personality and our vitality by making choices: these choices represent our power, and make us who we are. We pride ourselves on being self-reliant and liberated, the creators of our own destiny.

According to this story, every organisation, from businesses to government at every level, exists to meet our needs (or rather wants). We compete with one another to gain access to more of their bounty; they compete with one another to serve us, sell to us, and reinforce our all-important status. The twin competitions increase the range and raise the standard of the choices available. The higher we climb up society's ladder, the more dazzling choices are on offer for us. When each of us doggedly pursues our self-interest, that adds up to the best outcomes for society as a whole. Everybody wins, or at least, everybody who deserves to.

So the story goes.

To take a step back, there is of course a very basic sense in which humans have always been consumers: we have always breathed (consumed) air and eaten (consumed) food. For almost as long, we have been consumers in a further, more 'economic' sense: people have traded, bartered, bought and sold goods for as long as we have lived in groups. But all this was just consumption as one act among many performed by human beings. The Consumer Story only took shape much later, in the late 19th Century, and only became dominant across the world after the two World Wars, taking on its full form at the end of the last century and beginning of this. As the story took hold, consumption ceased to be just one act among many, and instead the why and the how and the what of every aspect of society came to follow Consumer logic.

Over the decades, this story has become so ingrained in our behaviours, institutions, and culture that we have ceased to see it. It is taken for granted, unexamined, unquestioned, invisible, like the air we breathe. It is further obscured by the fact that we think of ourselves as playing a multitude of roles – parents, students, employees, voters, shareholders, and so on, with consumer just one among many. Yet in fact the Consumer Story sits beneath, shaping how we fulfil all of them. We operate from within its parameters, without being aware of it, trapped. This is apparent when we describe the story's dynamics as 'human nature' and inevitable... just the way the world is.

Now the Consumer Story is collapsing in on itself, and is at risk of taking us down with it. It was supposed to set us free and give us power, but many of us feel trapped in a rat race, while the background to our lives is a spiral of decline we are powerless to do anything about. Its insistence on our narrow individuality is limiting us, holding us back and telling us we're not the ones to solve the problems of our world. Instead the story tells us that our agency is limited to choosing between options someone else provides. It has become unsustainable in every sense of the word. We have such pervasive extreme inequality it threatens the safety of everyone (even the wealthiest), while the story says that our primary responsibility is to compete against everyone else to hoard more. We have ecological breakdown, while the story insists that our identity and status rely upon further con-sumption. We have an epidemic of loneliness and mental health challenges, yet the story tells us we stand alone.

Our society is collapsing because the story we use to make sense of ourselves is collapsing. In order to step into a new story, we first need to learn to see the air we breathe, this story that shapes our world and behaviour, for what it is.

1984

The launch of the Apple Macintosh was far from the only telling moment of 1984, a year I have come to see as the zenith of the Consumer Story. The events of the year offer us a way to see both why the story held such allure and, with the benefit of hindsight, why it is now in the process of collapse. It is a tale of broken promises.

The Apple ad kicked the year off with the first promise: that the new wave of technology would empower us and save us from Orwell's nightmare. Technology was supposed to make us kings, with companies competing with one another to better meet our needs, making our lives easier and more convenient. Instead, as we now know, that nightmare was simply delayed. In the ensuing years, both states and corporations have gained the power to monitor our every move. We – in the form of our data – have become the products bought and sold by the tech giants, their revenues generated by trading our attention between themselves. Trapped in the story and so unable to see a way to act together to limit their excesses, we quietly accept their cookies and share our lives while their power grows and grows.

Two more Consumer superbrands arrived in 1984, each holding their own analogous promises. When Richard Branson's first Virgin Atlantic flight took off that summer, he was launching not only an airline but an ideal, almost a creed: the concept of customer service had arrived, and that great Consumer Story mantra, 'the customer is always right.' Branson would go on to build a business empire by bringing this ideal to bear on everything from banking to broadband, from hotels and holidays to gyms and health centres, explicitly choosing industries where we the Consumers were as yet poorest served. When the Virgin brand entered a new sector, its promise was to instigate – and win – a competition to raise standards and to innovate to serve Consumer 'needs' better and better. This was in turn part of the grand promise of which the Virgin Group has been a chief standard bearer, alongside brands like IKEA and Walmart: that when this kind of competition and innovation reached every sector, it would create a better world for all.

It hasn't worked out like that. Instead, our individual whims and wants have become fetishised, with ever more ludicrous business models evolving to meet and raise our Consumer desires and expectations. Many of us are working ever harder, while having ever more debt and fewer savings. What is more, this process has intensified even as our actual needs – for healthy ecosystems, coherent societies, resilient relationships – have not just gone unmet but been increasingly undermined. Such collective needs have been rendered invisible inside a story that values only the individual. As if to underline the breaking of that promise of a better world for all, the inaugural flight of Branson's Virgin Galactic in July 2021, 37 years after that of Virgin Atlantic, came in the middle of the world's hottest month since records began.

And so to the arrival of the third great superbrand of 1984. Nike sold its first Air Jordans at the end of that same year, and in doing so made its own leap from a technical sports shoe specialist to a brand fuelled by aspiration and celebrity. Nike's promise was to unleash the athlete in everyone, to give everyone access to the life of the stars. What we got instead was sweatshops and status symbols. Air Jordans quickly became and have since remained possessions that were not just desirable but essential to be accepted as a meaningful and valued member of society. The language of necessity here should not be dismissed: by one estimate, 1,200 young men are literally killed for their footwear every year in the United States alone.[75] And all the while, elite sport – the world Nike promised to open to us all – has instead become an ever higher paid, more exclusive and more distant dream.

There remain more broken promises to be found in that fateful year, adding texture to our understanding of the Consumer Story and its failure to deliver. Another came when The Body Shop founder Anita Roddick floated shares in her company on the London Stock Exchange at 95p, a price from which they rose rapidly, becoming known as 'the shares that defy gravity.' Roddick asked why people should have to buy all this plastic, and why we couldn't buy just as much as we wanted, like we could with fruit and vegetables. In doing so, she reminded us of our power as Consumers and even went so far as implying that our shopping could save the planet. The logic was that the greater the demand for The Body Shop's ethical products, the more other companies would follow the lead and offer their own 'good' products, possibly even leading government to raise the minimum standards around labour, the environment, and so on.

This vision of how to be good broadened at the end of 1984, when the biggest selling single in UK history, with six million copies sold and £8 million raised, became Band Aid's *Do They Know It's Christmas?* Bob Geldof, lead singer of British rock band

The Boomtown Rats, was moved by news of the Ethiopian famine and invited his pals, everyone from U2 to Phil Collins to George Michael, Sting, Paul Weller, Paul McCartney, to record a song. Geldof went on the radio to promote it, saying: 'You don't buy this because it is a good record. You buy this because it is your way of helping. Everyone has to do what they can... It's only £1.30. That's how cheap it is to give someone the ultimate Christmas gift ... the price of a life this year is a piece of plastic with a hole in the middle.'[76]

This beautiful promise - that all we really need to do is buy better stuff and get others to do the same, and any outstanding problems would be fixed - has sustained the Consumer Story far longer than it might otherwise have lasted. Certainly, for my part, it kept me working in the advertising industry, losing myself in selling trains over planes, or a toilet roll brand that promised to plant three trees for every one cut down. While there are some lovely brands out there, with people working really hard – like Tony's Chocolonely trying to end slavery in the chocolate supply chain or Oatly trying to eliminate the carbon emissions from dairy – the unfortunate truth is that this whole theory of change is fundamentally flawed.

The problem is rooted in the central falsehood that choice from a menu of options constitutes power. Ultimately, this serves to reinforce the idea that the limit of our agency as individuals is exactly that, as individuals, when the challenges we face are fundamentally collective, and require collective action in response. All that had really happened when The Body Shop hit the mainstream and Band Aid hit number one was that the Consumer

Story had eaten activism. For several decades, while the deeper problem was that collective power had been destroyed, many of us would be pacified by the Consumer conviction that we could shop our way to a better world.

THATCHER'S DREAM

To understand fully how collective power was broken, we need to turn to two final events of 1984, both coming in the month of November, and both involving then British Prime Minister Margaret Thatcher.

On 16th November 1984, three million shares in state-owned British Telecom were put on sale to the British public. 51% of the company was sold, just enough to give shareholders, rather than the state, the controlling stake. British Telecom had been 'privatised,' a term that had only been first coined in 1981, and a process that had never been carried through on anything like this scale anywhere in the world. For Margaret Thatcher, it was the ultimate expression of her vision for the country as Prime Minister. She believed herself to be on a moral mission, one entirely recognisable as the Consumer Story. She wanted to throw off the cloying hand of the state, and leave the people to climb the ladder unhindered. Her dream was to cement the meritocracy: anyone would be able rise to the top through their own efforts, with nothing to stand in their way but themselves. Privatisation was a key pillar of the plan: 'Through privatisation - particularly the kind of privatisation which leads to the widest possible share ownership by members of the public – the state's power is reduced and the power of the people enhanced.'[77] British Gas,

British Petroleum, the water companies, and the railways would follow. By the time of the 1992 election, the end of the Thatcher years, around two thirds of the UK's public industries, employing some 900,000 people, were no longer owned by the state but by private shareholders.

When privatisation came together with deregulation (epitomised by Thatcher's sweeping overnight deregulation of financial markets in the 'Big Bang' of 1986), the very idea of the state was remade to fit the Consumer Story. What could be provided for consumers by businesses would be, with government getting out of the way as much as possible. From inside the Consumer Story, this is the optimum mode. Why? Because people – Consumers – win, in several ways: we get better service from businesses, with competition driving quality up and prices down; we get better service from government, with the state not only gaining the revenue from selling off state-owned industries but also concentrating on what it should rightfully do; and we even get more power, as some Consumers become company shareholders.

What was really happening here, though, becomes clearer in light of the other seismic moment of November 1984. Just eight days before British Telecom's shares went on sale, the beginning of the end of the biggest labour strike in world history was signalled with the first miner's return to work. The strongest union in Britain, the National Union of Mineworkers (NUM), had gone on strike in early 1984. Eight months later, over 26 million person-days of work had been lost. The then primary means of energy production of one of the world's most powerful nations was in effect suspended. It was an arm wrestle of the highest order. Mines might have been owned by the state and so theoretically controlled by the government (through the National Coal Board), but there was little doubt the NUM was really in charge. It had won its previous battle with government in 1972, winning a 27% pay increase for its members and dealing what many historians consider a fatal blow to the authority of then Prime Minister Ted Heath.

But its opponent now was Thatcher, Heath's successor as Leader of the Conservatives, Prime Minister since 1979, and fresh off another election victory in 1983. She was ready for a fight. As oil, gas, nuclear and renewables became increasingly viable as alternative power sources, and the first whispers of climate change spread, the government quietly built up its coal stocks and began to close mines, triggering the strike pretty much deliberately in the process. Thatcher and her ministers had the story they wanted to tell ready to go. In an early interview, Peter Walker, Secretary of State for Energy, framed the activities of the NUM as 'a challenge to British democracy and hence to the British people.'[78]

When that first miner returned to work, it would still be a few months before defeat was made official, but from that moment on, it was coming; and when it came, it was complete. Miners across the country went back to work with no agreement whatsoever in place; the government could in effect do whatever it wanted. The consequences would be felt far beyond the mines. Mick McGahey, NUM Vice President, would later reflect that the result of the strike was 'to destroy trade unionism not only in mining but in Britain.'[79]

As Thatcher saw and encouraged the nation to see it, an over-powerful and under-accountable union was throwing its weight around, flooding the market with an unviable technology at inflated prices, not only costing the British people (British Consumers) money, but also depriving them of choice. She was not entirely wrong. Yet the strength of her Consumer-centred conviction rendered virtually invisible some important and inconvenient facts: that miners were people too; that industrial towns across the country depended on mining as the anchor of their local economies and communities; and that the government offered almost nothing to support those communities through the transition to new sources of power. It was not necessarily wrong to take on the NUM, certainly not wrong to start to move Britain beyond coal – but the way in which it was done was driven by the most cut-throat version of the Consumer Story. It created untold damage, the legacy of which underpins many of the divisions in Britain today.

The reputation and perception of unions as a form of collective action was one particularly important casualty of this moment. At their best, unions have through history provided a vital check and balance, representing the collective interests of workers and producers, 'the people,' in constructive tension with business owners. The Consumer Story moved the focal point. Now 'the people' were the Consumers, not the producers or workers, and business owners and the state took the people's side; making the unions the enemy. With their leaders often displaying the same blockheadedness as NUM President Arthur Scargill in 1984, that reframing has never truly been undone, and unions have been largely frowned upon in Britain ever since. The lack of meaningful pressure to regulate and reform the gig economy operators in this country – Uber, Airbnb, and the like – is just one of many symptoms.

When combined with privatisation, however, the true significance of the events of November 1984 is greater still. This was the moment of the fundamental breaking of collective power by Consumer power. Unions and the state might have been outdated, but nothing was nurtured to replace them. Instead, all that was left was individual, atomised power: power that could only be mobilised by appeal to self-interest, and could only be expressed in the choice between options, not the power to shape those options.

Our world today is entirely recognisable as the legacy of 1984. As Thatcher saw it, the previously pervasive reach of the state – from schools to housing to industries like telecommunications and mining – encouraged dependence and helplessness, inhibited competition, and deprived people of the opportunity for self-determination, the power of choice, and the necessity of taking personal responsibility for their actions and their outcomes. When she said 'people,' she was picturing self-reliant, independent, atomised individuals – or at most nuclear families – for whom the state should merely provide options to choose from. We could, should, and would stand or fall by our choices. Many of us, of course, would fall.

Today, many of the biggest companies are both bigger and more powerful than even the biggest states, and we have come to see them as the truest arbiters of human will. There has been an overwhelming transfer of power and money from elected governments to shareholders, justified on the basis that this is a greater and more democratic form of power – yet those shareholders are still a wealthy elite. A default assumption has developed against market interventions of any sort, not just to the extent of state-owned industry but all forms of regulation. There have been successive reductions in the status and funding of local councils, as they have been reduced to branch outlets of the national service provider that the central government has become. Unions have been, if not completely destroyed, then certainly denuded of their relevance. Even the basics of democracy have been affected by the expansion of the Consumer Story, with the role of the individual limited to voting at most, and that on the basis of a choice as to which of the options available will provide the best service.

THE RISE OF THE CONSUMER

The Consumer Story did not arrive in the world fully formed; it was not conceived, developed, and then imposed on society by some mastermind. Instead, it took shape a little at a time, gaining coherence, momentum, and reach along the way.

The word 'consumer' appears as a noun for the first time in the work of British economist William Stanley Jevons in the late 19th Century: in articulating 'use value' he was the first to argue that the value of, say, a watch depended most importantly not on what it cost the producer to make it, but on how much 'the consumer' wanted it.[80] The verb 'consume' had long been in use, but with this linguistic step, Jevons for the first time created 'the consumer' as an identity construct. Despite this original articulation, the Consumer Story did not take on its full shape in Britain. It needed a more welcoming culture in which to grow. Britain's ingrained class structures and Victorian morality were inhospitable; instead, as the 20th Century dawned, the new story would grow to maturity across the Atlantic, in the land of liberty: the United States of America.

In declaring independence from Britain, the Founding Fathers rejected its monarchy and aristocracy, and the centuries-old systems of patronage and inheritance. America, by contrast, would be ruled by professionals who earned their place at the top. America's fabled frontiers offered freedom, limitless growth, and abundance to its rugged individualists and lone rangers, not just geographically but economically and socially. Limits and scarcity were Old World ideas. Liberation, independence, potential, self-reliance, choice: these constituted the American Dream. They were also the chief selling points of the Consumer Story.

By the 1920s, the story was still not globally dominant but robust, nearly ready for its ascendance. This was the era when Edward Bernays, nephew of Sigmund Freud, took his uncle's theories about what drives human behaviour, the subconscious and unconscious, and created the field of public relations and advertising in its modern sense. In the aftermath of the horrors of World War I, the increasingly influential Freud had come to see human nature as inherently bad, believing that we have an innate, instinctive desire to kill, for example, and an insatiable and deeply nefarious desire for sexual gratification. For Freud, government was a necessity to protect us from the chaos that inevitably ensued when our true selves came out. Bernays would inherit his uncle's beliefs, and use them to justify his own hero complex. Consumption was the best possible release valve for humanity's pent up nastiness, the way to save us from ourselves. 'Engineering consent' was the social responsibility of people like Bernays and his fellow public relations professionals.[81]

Bernays' first great success was for the American Tobacco Company. He was tasked with overcoming the resistance among women to buying and smoking cigarettes. At the time wider society judged women smokers to be immoral and even criminal. Bernays needed to reframe the conversation. He achieved this by infiltrating the annual Easter Parade on 5th Avenue in New York City in 1929 with a crew of fashionable young women – the influencers of the day – who flaunted their cigarettes as 'torches of freedom.' Bernays successfully recast this act of consumption – smoking – as an expression of defiance and liberation, chiming with the era's rising demand for equality for women. It was a perfect example of how identity could be created and expressed through consumption. Bernays would go on to advise multiple administrations, viewing politics as just another product to be sold. As the Third Reich's Minister of Propaganda, fanboy Joseph Goebbels applied lessons learned from Bernays to create the 'Führer cult' around Adolph Hitler.

When World War II ended, the Consumer Story quickly became America's greatest export. This is in some ways a statement of the obvious: it was the postwar decades that we all associate with the proliferation of consumer goods and services. In the United States, consumption was explicitly equated with patriotism and democracy, and Americans dutifully purchased millions of refrigerators and freezers, vacuum cleaners, washing machines, cars, and the rest. With America's new global superpower status, the story spread, and consumption grew across the world. Although there were large parts of the world where lower incomes restricted American- and Western-style levels of consumption, the aspiration – which is to say the underlying story – was still taking root.

What was most significant about this time, though, was not the increase in material consumption, but the new structures and institutions that were created in the process of remaking the world. First came the World Bank and the International Monetary Fund (IMF), conceived at a meeting of 43 nations at Bretton Woods in 1944. The European Coal and Steel Community (ECSC), the precursor to the European Union, was formally established by the Treaty of Paris in 1951. The Organisation for Economic Co-operation and Development (OECD) followed, again in Paris, in 1960. All of these were created to promote trade – in the case of the World Bank/IMF, to keep currencies stable or provide sufficient resources to make purchases – on the hypothesis that trade was the way to sustain peace. Trade would promote consumption, and consumption, per Freud, would harmlessly channel energies that would otherwise only cause trouble. As French Foreign Minister Robert Schuman put it in the speech that launched the ECSC, the explicit intention was to 'make war not only unthinkable but materially impossible.'[82]

What these first institutions represented, for better or worse, was the Consumer Story taking on tangible shape and seemingly permanent structure. This was a critical step. Once the story had expression in global organisations and institutions, we were committed. With the institutions came new goals and measures that would carry the story further into society: it was at Bretton Woods, for example, that Gross Domestic Product (GDP) was adopted as the principal measure of the success of a nation, and thereby became the focus of political attention. From this point on, growth was not just good, but the primary aim of political and social systems. We all know now where that has taken us, but some saw it early, including the economist and retail analyst Victor Lebow in 1955:

Our enormously productive economy demands that we make consumption our way of life, that we convert the buying and use of goods into rituals, that we seek our spiritual satisfactions, our ego satisfactions, in consumption. The measure of social status, of social acceptance, of prestige, is now to be found in our consumptive patterns. The very meaning and significance of our lives today expressed in consumptive terms. We need things consumed, burned up, worn out, replaced, and discarded at an ever-increasing pace. We need to have people eat, drink, dress, ride, live, with ever more complicated and, therefore, constantly more expensive consumption.[83]

Over the next decades, that is exactly what happened. Consumption expanded and became more sophisticated, with the spread of the Consumer Story driving it, sprouting more new institutions, structures, and measures along the way. In 1967, for example, a US business association launched a new measure called the Consumer Confidence Index, asking a nationally representative sample of the population on a regular basis to report their intention to spend more in the coming months than in those just past. It proved a powerful proxy indicator for GDP, and expressed the story – and the contribution they could make – in a way that people could understand. Unsurprisingly, it spread across the world. As it did so, like GDP, it became not just a measure of society, but a moulder of society: politicians and civil servants aiming at increasing GDP began to aim at increasing Consumer Confidence, and to think of people more and more as Consumers – explicitly, uncritically, and exclusively.

With the stage set, popular culture played along through the 60s, 70s and into the 80s, with songs, TV shows, and films spreading the gospel of the Consumer: idolising the individual in the form of heroes and stars, portraying success in purely material terms, glorifying competition – all of it punctuated by the steady drumbeat of advertisements that sold us the accoutrements we needed to embark on our own heroic adventure, or on the items that would conspicuously signal our success and our aspirations to others. Our prison was complete, to the extent that we had set about decorating it for ourselves.

LIFE INSIDE STORIES

I was brought up by parents who lived the post-War dream, each in their own way. Born in 1940, my Dad was among the first to benefit from the emerging meritocracy, plucked from his small town, home counties, steam trainspotting childhood and escalated into Shrewsbury School on an academic scholarship in the late 1950s. He would become a civil engineer, building the infra-structure (including his beloved railways) of the post-War world, from South Africa to the Middle East and, later, post-Communist Poland.

My mum, 12 years younger and driven by a tenacity and defiance she inherited from her forebears, came from lower stock in a way that only became possible in the context of war, and flourished in a way that women of previous generations could barely have imagined. Her uncle Alan rose from whipping boy to Wing Commander in the Royal Air Force; Auntie Gwen was among the leaders of the women's working revolution back home. Mum would work as a secretary and interpreter in Geneva, set up her own business, and raise and inspire my brother and me.

Their lives are examples of freedom powered by consumption, the Consumer Story at its best. The arrival of personal computing and international travel transformed my dad's industry and career; my mum remembers her family's first washing machine, a Hotpoint Liberator, emphatically deserving its name. It's easy to see why I went to work in advertising. It's obvious why humanity got trapped in the Consumer Story. We have been conditioned to believe that, fundamentally, we are Consumers. The story that promised our liberation has become our prison.

This comes as no massive surprise. Humans get trapped inside stories all the time, both individually and collectively, in small groups and on a massive scale. As babies, for example, how our parents or caregivers treat us imprints us with an 'attachment style' that governs how we relate to others for the rest of our life.

Their feedback, verbal and nonverbal, deposits the first layers of sediments in what become our 'limiting beliefs' about ourselves and the world, ideas that get further reinforced by schooling and the world at large. So long as these stories remain unexamined and unquestioned, they are as limiting to us as the iron bars of a cell.

Tell us a story that the earth is flat, and we'll refuse to set foot on a boat for fear we'll fall off the edge. Tell us a story that women are weaker and more emotional, and we'll pay them less and deny them the vote. Tell us a story that God demands it, and we'll capture, convert, and exterminate on His behalf.

Indeed, almost all the inviolable, incontrovertible-seeming assumptions of our society can be understood not as inevitable facts of life, but as stories. Even in the realm of science, theories are laws only until they are replaced by the next theory, or story. It is just a matter of time. For centuries, the story was that the Earth was the centre of the Universe, and that all the stars and celestial bodies revolved around us. This belief had implications not just for the physics of planetary motion, but for our spirituality, and for how we structured society. When Copernicus and his peers challenged that story, they were ridiculed and ignored, because people were so invested in the idea of the Earth as the centre of the Universe and so frightened about the implications of the new story. When there was finally consensus about a heliocentric worldview, the ripples extended to all sorts of scientific disciplines and to how we thought of ourselves.

One set of people at one moment in time might consciously agree to such a story, but it goes on to have a life of its own over subsequent generations, informing and inflecting not just our individual lives, but also organisations and institutions, and ul-timately, our entire culture and society. We live in the legacy of stories created generations ago. All of us inevitably perpetuate stories; no one, really, is to blame. Once a story is widely held and deeply rooted, this in turn shapes all further institutional and individual possibilities and actions. When it's time for the people in charge, or the group or institution or society, to make a decision, particularly in a stressful context with potentially dire consequences, the default or dominant story or strategy is gener-ally the only option they can see. They become like the drunk who searches for his lost keys solely under the streetlight because that is where the light is.

Stories function like the streetlight in a moral as well as a prac-tical sense. The light illuminates those who are good, who count, who should be thought of as 'us'; it leaves in the darkness those who are bad, who don't matter, who can be dismissed as 'them.' American linguist and political scientist George Lakoff calls this effect 'framing,' and has devoted his career to understanding how metaphors (stories) can exert a huge influence on political deci-sions. Among many other contributions, Lakoff famously set out 'the fairy tale of the just war': 'A crime is committed by the villain against an innocent victim... The villain is inherently evil, perhaps even a monster, and thus reasoning with him is out of the question. The hero is left with no choice but to engage the villain in battle.'[84] By defining who is the victim, who the villain, and who the hero, political leaders are able to tell stories that take a nation to war.

This takes us back to our search for an explanation as to how we can hold onto the idea that Citizenship is human nature, while still having a clear-eyed view of the cruelties and catastrophes of our world. In almost every case, people do want to be good, but all too often we embrace a definition of goodness presented by someone who wants to benefit – politically, economically, socially. Think of the man who agreed to administer electric shocks in the Milgram experiment once persuaded that he was contributing to medical science. Stories have moral power, and it is a moral power that can be abused.

To take everyone's favourite example of evil: between 1933 and 1937, Hitler spun a powerful story in both actions and in words. It was a story about giving Germany back its pride as a nation: he would dismantle old entrenched class structures and systems and raise up the millions who were left poor and destitute by the economic humbling of the nation after World War I. In a time when inflation caused a loaf of bread to cost a million marks, Hitler introduced the Volkswagen (literally 'the people's car'), the creation of traffic-free inner city areas with strict pollution control, and the expansion of public green space. He literally provided a breath of fresh air to a population in pain.

Hitler's story said to his countrymen that to be good was to work hard and do your duty, and to trust in him to protect them: to get the wealthy and the outsiders, and particularly wealthy outsiders (i.e. Jews), out of the way. It said that he was pursuing the good of the German nation, and indeed that if other nations followed the same path, that would be good for the world. Not only did vast swathes of the German population buy into this story, so did plenty of notable foreign observers. Writing about that period in her deep study of the life and role of Hitler's fabled architect Albert Speer, historian Gitta Sereny describes it as 'the time most people now like to forget, when American expatriate writer Gertrude Stein thought Hitler should get the Nobel Peace Prize; George Bernard Shaw passionately defended him; the Swedish explorer Sven Hedin lauded his "indomitable passion for justice, breadth of political vision, unerring foresight and a genuine solicitude for the welfare of his fellow citizens"; and Britons such as Lord Halifax and David Lloyd George conferred their stamp of approval.'[85]

We recognise the similarities with the United States in the lead-up to the 2016 election of Donald Trump. The resonance of his story is powerfully explained by one of those who saw these results coming, the sociologist Arlie Russell Hochschild. In 2011, already concerned by increasing polarisation, she set out to bring her research methods to bear on the political divides in her own country. The resulting book *Strangers in Their Own Land* was published in 2016, before Trump's election to the presidency, and is the result of a five-year immersion in Louisiana searching for greater understanding. She pieced together a kind of collective story from the strands of the individual stories she was repeatedly hearing in her research:

You are waiting in line, working hard and doing the right thing in the expectation of future reward.

Suddenly, others – people who are not like you - are cutting in ahead of you. In fact, they are allowed and even encouraged to do so by government. You feel betrayed, frustrated, angry, resentful.

This was the story that the Tea Party members in Louisiana – predominantly low-income, working class, white, libertarian-conservatives – were telling themselves about their lives, and about the threats that immigrants and other Others posed to them. Hochschild called it the 'deep story' of that community, describing it as the 'subjective lens' through which they saw the world.[86] With 'Make America Great Again,' the Trump campaign told a story that built directly onto this deep story, promising to fix the situation, and that would enable Americans to be the good, hard-working people they wanted to be. Hochschild also recognised that deep stories come in interlocking layers, some deeper than others, with the deeper dictating the available space for the shallower: personal deep stories exist within community deep stories exist within national deep stories.

Hochschild's ideas have fundamentally shaped my understanding of the stories of Subject, Consumer, and Citizen. As I see it, they represent a kind of deep story that occupies a foundational realm, even deeper and lying underneath the others that Hochschild observed, informing and inflecting everything we can imagine, everything we build, how we act and interact. This kind of story is common to all members of society, and shapes every aspect of the interplay of self with society. It exists everywhere, though it is strongest and purest where global power is centred. It is constantly reinforced, which is ultimately what those 3,000 commercial messages a day are doing. Indeed, even challenges to it often reinforce it still further, in the same way that language like 'post-' and 'anti-' don't offer alternatives, but just maintain the dominant story even in their rejections of it.

This kind of story sits deeper than any piece of literature or propaganda, deeper than our epic myths or religious texts. It operates at the level of world-shaping that Donella Meadows, one of the pioneers of systems thinking, articulated as the paradigm, which she described as: 'The shared idea in the minds of society, the great big unstated assumptions – unstated because unnecessary to state; everyone already knows them – ... the deepest set of beliefs about how the world works...'[87]

When we come to see this for the first time, it can seem overwhelming. The Consumer Story is so foundational that changing it requires a massive shift. Yet then again, the very word 'consumer' didn't exist just 150 years ago. The story has only held global dominance for one human lifetime. That means stories can change.

4. ONCE WE WERE SUBJECTS

Before the Consumer, there was another story in place: the Subject, as in 'subjects of the king.' In this story, the Great Man – the Chief, Pope, King, Boss, Father – knows best. The rest of us are innocents, ignorant of important matters. We must rely on him to chart the way forward and declare our duties. Our part is to obey and accept what we are given. In return, he will protect us and maintain order, a deal that is more attractive the greater the danger. Governments and organisations that arise out of the Subject Story are paternalistic and hierarchical, with the inherently superior few at the top of the pyramid.

In light of this story coming before, it is little wonder that the Consumer felt like a breath of fresh air. But the Subject Story does not only make us feel better about having fallen for the Consumer; its very existence is evidence that stories can and do change. Deep stories feel inevitable once they become entrenched, but they can shift. Indeed, while the Consumer Story has had a long enough hold, the Subject is much, much older. It was the dominant story across the world for centuries, shaping the interactions of the majority of humanity with our institutions and each other from at least the 1600s; but it's not hard to find evidence of it going back millennia.

We can see early traces of the Subject Story as significant numbers of people stopped hunting and gathering and instead took up the cultivation of plants and animals. Agriculture changed the way humans lived, bringing us into closer proximity with each other as we settled to tend crops and 'livestock.' This in turn affected our social structure. Animal husbandry and grains ensured year-round provisions, but at the same time meant there was more to lose. Wild animals, the people of other settlements, and the large proportion of the human population who maintained the hunting and gathering lifestyle often saw fit to help themselves to some precious crops or stores of harvests. In response, the first settlements evolved into the first states, and the first fences and soon walls were built around them.[88]

Those first walls represented physical protection and security, key dynamics of the Subject Story. Physical walls were also mental walls: they were the first great act of separation, manifesting the idea that some should be in and others must be out. This was the moment humanity first separated from nature in order to subjugate it, as is implicit in that word 'livestock': those living creatures who came inside the walls were now reimagined as material possessions. As part of the same dynamic, this was also the moment when the crucial, mysterious power of the female animal to propagate life became an essential focus. The human female was no exception: women started to be safeguarded and controlled, as the foundations of patriarchy were laid.

The walls split humanity itself into two fundamentally distinct groups: hunter-gatherers and settlers. Hunting and gathering might have remained the majority mode of life across the world until as recently as the 1600s, but those who pursued it have never held the pen of history; it was those inside the walls who would dictate the terms. Looking through the eyes of those early settlers, we can perhaps imagine how they might, with encouragement, have come to see hunter-gatherers as more like animals than like themselves; how the difference they perceived (and the threat they felt) might then have been translated into self-declared superiority. Separation bred superiority bred dehumanisation, and it started a long, long time ago.

Around the time of those first walls, likely preceding them in some places and following in others, came three further innovations: kings, writing, and organised religion. It was when all these came together that the settlers explicitly became the first subjects, as in 'of the king.' That had happened by 2330 BCE in the figure of King Sargon of Akkad, who became ruler of the world's first known empire as he established control of a network of city-states across ancient Mesopotamia, crushing the power of his rivals, enslaving opponents, and installing puppet governors as he went.[89]

Systems of writing had evolved in the same region several centuries before, initially to keep tabs on what everyone was producing and what amount of tax they owed as a result. Sargon made full use, spending the taxes he collected to pay not only his soldiers but also an array of royal artists and scribes. These documented and glorified his wisdom and his role in providing order and dispensing justice – prefiguring the legal codes that would soon follow – as well as his military prowess, and elevated him to the status of a demi-god in the process.

Sargon's reign, over four thousand years ago, set the archetype of 'king as father.' He had himself portrayed as paternalism embodied, the Great Man, an ideal leader who stood above his people and knew what was best for them. In order to do so, Sargon also invented patronage as a business model for the arts, recognising the capacity of culture to embed his power and expand it even beyond his lifetime; and at the same time cultivated organised religion, framing his power as derived from the gods, himself as their interpreter and emissary. Yet of course, for all this paternal wisdom, for all the cultural legacy, Sargon's reign was built – as the Subject Story ultimately always is – on raw power.

We still know only fragments about the structures of the societies over which kings like Sargon ruled – but the Code of Hammurabi, a set of 282 laws carved onto a single black stone pillar some time between 1792 and 1750 BCE, fills in many of the blanks. What emerges by implication, from a set of rules governing everything from family life to professional contracts to administrative law, is that the life of individuals was rigidly structured and stratified. There were very different standards of justice on offer depending on which of three classes one was a member of: property-owning men, freedmen, or slaves. A woman was considered the property of her father or husband. Social order comes through as infinitely more important than individual rights. The overall impression is very much of a society where people accept their station, do their bit, and obey the rules and the King from whom those rules came.

Contrary to the old idea of the Agricultural Revolution as a sudden, joyous respite from a life 'red in tooth and claw,' this story was neither willingly nor swiftly embraced. Recent developments in archaeology have led many to argue that humanity had to be dragged kicking and screaming into permanent settlement, that the first walls were often as much to keep people in as out. Once imposed, though, the lesson of Sargon is that from very early on there was not just physical safety on offer to sustain the story, but also order, and perhaps most compelling, a clear and highly addictive sense of 'Us and Them,' an allegiance to something bigger than self from which to draw meaning in life.

From this point on, the core dynamics of the Subject Story were largely set: protection, division, subjugation, conquest, patronage, duty; in a word, paternalism. We can recognise them from the Roman Empire to the British to the rise of Hitler, right through to the looming resurgence of the Subject Story in the form of the strongman leaders of the present day. Over the centuries that followed, the story emerged along similar lines in locations across the world – Egypt, Mesopotamia, the Indus Valley, China, the early Maya – and from these epicentres spread and scaled, increasing in size and sophistication but without altering in the core dynamics. Confucius, around 500 BCE, taught that the ideal ruler was a 'superior man' – *junzi* – who was required to model the morality he wanted his subjects to cultivate; a century or so later, Plato made a similar case for 'philosopher kings' in his *Republic*; both were articulating much the same ideal that Sargon had embodied nearly two thousand years earlier. And from those first walls in our minds, others followed, more sturdily built: Aristotle, student of Plato and tutor to the Macedonian emperor Alexander the Great, put forward 'scientific' arguments for the existence of a 'hierarchy of souls,' which ratified the superiority of humans over animals, men over women, and some races over others, who should be considered 'natural slaves.'

The spread of Christianity echoed the principles of the Subject, putting forth the ultimate Father, the almighty God in whose great hands lay the fate of all His people, the shepherd to us as flock. Under Him came another father-figure in the form of the Pope, under whom came widely distributed Fathers of local churches. A veritable Russian doll of paternalism. Later, despite the ostensible democratisation of faith and anti-corruption stance that the rise of Protestantism promised, the reformed religions would crystallise a self-sacrificing work ethic and minimalist lifestyle that conveniently venerated the Subject condition: work ourselves to the bone and shun all earthly comforts, and we too can earn our way to a place in the Father's heart in heaven.

Across the known world and in all its guises, the expansion of the Subject Story was interrupted by regular rebellions and uprisings on the part of subjected peoples, who were understand-ably unwilling to accept the inferiority declared for them – the Subject Story was a hard sell throughout, not just at the beginning – but for the most part these were either mercilessly crushed, or co-opted with minor concessions that served more to increase the sophistication of the story than change it fundamentally.

By the late 15th Century CE, the story had reached the limits of its expansion on the European continent. The ongoing fights were now almost exclusively about who was subject to whose rule, rather than the expansion of the Subject Story itself. Hunter-gatherers were by now confined to the margins, where states could not function; although out of reach of written history, many humans continued as they had for aeons. Then came the 'Age of Discovery.'

Over the next four centuries, the Subject would become the dominant Story across the whole world, with ever more power concentrating in ever fewer centres, almost all of them in the European states. They would be spurred on in their conquests by competition with (and fear of) one another, and the moral mission of saving the heathens which was explicitly spelled out by the Church. Edicts from the Pope gave Christian explorers the right to claim lands they 'discovered' and lay claim to those lands for their Christian Monarchs. Any land that was not inhabited by Christians was fair game. If the 'pagan' inhabitants could be converted, they might be spared. If not, they could be enslaved or killed.

The British, French, Dutch, Belgians, Italians, Russians, Portuguese, even the Danes and Swedes claimed swathes of territory during this era, enabled by the Subject lens through which they saw the world to ignore completely the fact that – contrary to the idea that they were 'discovering' these 'new' lands – there were plenty of other humans out there. The power of the Subject Story helps explain how this could have happened. In each and every case, the colonisers fundamentally failed to recognise the shared humanity of those whose lands they claimed.

And so to the British Empire, the culmination of the Subject Story and the epicentre of its collapse. If the United States of America is the nation that epitomises the Consumer, Britain – or rather the British Empire – is the equivalent for the Subject.

1897

At Queen Victoria's Diamond Jubilee on June 22nd 1897, the Empire over which she ruled was the largest ever known, spanning a quarter of the land surface of the planet and encompassing nearly as great a proportion of its people; British colonies and 'protectorates' ranged from Hong Kong and Singapore to Canada and Australia, and included the whole of the Indian subcontinent. The pomp and circumstance of that day were unrivalled, with the Empire brought together to parade the streets of London in Victoria's honour. Three million spectators thronged a city dressed in red cloth, flowers, and triumphal arches. It was 'the brassiest show on earth,' in the words of historian Jan Morris: 'There were Rajput princes and Dyak headhunters, there were strapping troopers from Australia. Cypriots wore fezzes, Chinese wore conical straw hats. English gentlemen rode by, with virile moustaches and steel-blue eyes, and Indian lancers jangled past in resplendent crimson jerkins.'[90] Victoria sent a message that day that was telegraphed to quite literally every outpost of the Empire: 'From my heart I thank my beloved people,' she said. 'May God bless them.'

It is a moment that evokes so much of the legacy both of the British Empire and of the Subject Story as a whole. It conveys the pride, the sense of duty and shared allegiance conjured among so many. My countrypeople love to serve a cause, to call their homes their castles, to wave a flag to 'Rule Britannia' and 'Land of Hope and Glory.'

And yet at one and the same time – and from only a slightly altered perspective – we can also see all too clearly the horrendous sense of superiority, the dehumanisation and exploitation that drove it all: the nation itself as King, the rest of the world its Subjects. This is a legacy the British have yet to face up to in any meaningful way. A recent survey found that a third of my fellow Brits view the Empire as something to be proud of, with fewer than a fifth seeing it more as cause for shame; a third again believe the nations colonised were better off as a result; more than a quarter wish Britain still had an Empire.[91] The sentiment is present in other former imperial nations. Our failure to confront what happened and what continues to happen represents a continued abuse of the nations and peoples we conquered and exploited – as British, as Europeans, as white people. It also constitutes a delusion even in the limited terms of narrow self-interest: until the image of the glorious past is challenged, it's easy for many to hanker after a return to a mythical golden age, when in reality almost all were exploited in the time of Empire, those abroad just rather more than those at home.

By nature a proud Brit myself, I count myself among those who have found this a difficult legacy to face. I have, however, been able to hold it with a little more humility and a little less fragility by viewing it not through the lens of national exceptionalism, but as the last major instalment – the climactic culmination perhaps, but still only one among many – of the Subject Story.

As I see it, the core driver both of the British Empire's final expansion and of its collapse was one and the same: the Industrial Revolution of the late 18th and early 19th Century.

Industrialisation took hold first in Britain partly because of the existing extent and economic power of the Empire by that point, and served in many ways to consolidate and expand that power still further, certainly on the global stage. New technologies enhanced production at home and in parts of Imperial India to the extent that Britain was able not just to ride out American independence but even to expand into the Pacific in response, with the colonisation first of Australia and later New Zealand. Yet as time went on, the societal shifts caused by the Industrial Revolution began the unravelling of the Subject Story, and created the conditions for the emergence of the Consumer Story that would soon come to challenge it.

The new industrial middle class was a direct threat to the aristocracy that had for millennia been at the core of the Subject Story. For most of the 19th Century, the new businessmen and the old landowners co-existed happily enough: 'the business of businessmen was business; the business of landowners was government,' as one historian put it.[92] But over time, with cheap goods coming in from the colonies, and the urban industrial economy exploding, the rural economy – the wealth base and therefore the power base of the landowning aristocracy – began to show signs of weakness.

The landowners' response only made things worse. They increased domestic exploitation through a massive final wave of land enclosure, in the name of agricultural efficiency. Before enclosure, land was considered the possession of the owner only during the growing season: after harvest, the land was open to the community to allow grazing and gleaning on it. Enclosure made it the year-round possession of the aristocracy. Those who lived directly off the land were conscripted into the industrial workforce or reduced to beggar status.

Homelessness became hyper-criminalised, with many shipped off to the new penal colonies in Australia. Little was achieved for the agricultural economy; more significant was the extent to which this more intense subjection led the most exploited to ask more and more serious questions about the validity of the rules and the wisdom of their alleged betters. At the same time, the businessmen began to seek their turn at political power, feeling they could perhaps do rather better at government, not just wealth creation.

As the grip of the British aristocrats loosened from the national levers of power, so on the international stage, the grip of the paternalistic imperial nations would soon loosen over their Subject nations. Peasant revolts all over the European continent found an echo in unrest and resistance in the colonies and protectorates. The conditions for the ascendancy of the Consumer Story were beginning to develop, and not just in America. In India, three men would strike a blow at the heart of the idea and legitimacy of Empire and the Subject Story: Lala Lajpat Rai, Bal Gangadhar Tilak, and Bipin Chandra Pal. Although each came from a different part of India, differed in caste and religion, and disagreed on what kind of nation India should become, they were united on

India's need for independence from British rule. Once they united forces, they became collectively famous as Lal Bal Pal. In 1905 Lord Curzon, the British statesman serving as Viceroy of India, decided for purportedly administrative reasons to partition Pal's home state of Bengal, inflaming tensions between the Muslim majority in the east and Hindu majority in the west. Lal Bal Pal saw it as a strategy of further suppression, divide-and-conquer. In response, they organised a rebellion, using a tactic they called *swadeshi*, which translates from the Sanskrit as 'of one's own country.' It was to all intents and purposes a Consumer boycott: Indians avoiding British products in favour of Indian ones.

Britain's economic power depended largely on its factories. They imported raw materials like cotton from the colonies and exported finished products like clothing, including back to their nations of origin. The cost efficiency of such industrial methods meant that products arriving back on India's shores were often cheaper than those produced at home, undermining the development of India's own industry. *Swadeshi* inflicted economic pain at the same time as giving Indians an understanding of the power they had in their own hands; Gandhi would later take the idea and make it his winning strategy, with his homespun cotton and his Salt March.

Swadeshi epitomises the situation at the turn of the century. The new Consumer Story was now taking shape everywhere. The Subject Story was still dominant, but straining to hold. The idea that the right thing to do was to accept our station in life was finally coming under widespread question. In that context, Lal Bal Pal created a strategy born of the Consumer in order to attack the Subject. Not only was *swadeshi* rooted in the act of consumption, more importantly, it was a strategy of self: self-reliance, self-help, self-determination, on a national scale. It posited that the destiny of Indians was in their own hands, and that in acting in their own interest, they could change the very structures of society. This was unmistakably Consumer logic. There would be one hell of a fight though – or more accurately two – before the Subject Story would give way.

THE STORY WARS

History, as the saying goes, is written by the winners. This was certainly the case with the initial – and until recently, the default – understanding of the causes of World War I, which essentially puts the cause down to belligerence on the part of Germany and of the old Habsburg Empire of Austria-Hungary. The latter bullied its Serbian subjects in particular to distraction, with the more or less inevitable consequence of rising Serb nationalism. When Archduke Franz Ferdinand, heir to the throne, and his wife were assassinated by a Bosnian Serb in Sarajevo, an equally aggressive Germany pledged support to the Habsburgs. Austria-Hungary issued an ultimatum to the Serbs that was so harsh that it was little more than a pretence at peace. Germany invaded France via Belgium to the West, and confronted Russia in the East. In this way, to paraphrase the famous words of British Foreign Secretary Sir Edward Grey, the lights went out all over Europe.

In 2013, though, Christopher Clark, Professor of Modern History at the University of Cambridge, put forward an altogether different view in his widely celebrated book *The Sleepwalkers*.

For Clark, the picture was far messier. Trouble had been brewing across the continent for decades, as the structures of power in every nation from Victorian Britain to Wilhelm's Germany to Tsarist Russia became less certain, and those in positions of power became fraught and paranoid. Foreign ministers and monarchs were feeling intimidated and vulnerable as a result of events at home, new demands for accountability to their people, and desire for power on behalf of those who had made their money not from land but from industry. As a result, they looked abroad with increasing fear and suspicion. This created what Clark called 'a crisis of masculinity,' with Grey one of those most susceptible: 'competition from subordinate and marginalised masculinities – proletarian and non-white, for example – accentuated the expression of 'true masculinity' within the elites... it seems clear that a code of behaviour founded in a preference for unyielding forcefulness... was likely to accentuate the potential for conflict.'[93] Clark chose a softer metaphor for his title, but his argument was essentially that the leaders of the old world powers were like posturing stags, spoiling for a fight to prove their virility. The threat of the impending collapse of the Subject Story undermined these people, and rendered the continent tinder-dry. Ferdinand's assassination was just the spark that set it alight.

The nature of the fighting in World War I would make the Subject Story all the more difficult to sustain, and should have brought it to its final end. Flung together in the trenches, officers and infantry were viscerally confronted with their basic shared humanity in a way that made belief in the supposedly God-given hierarchies difficult to sustain. So did the recognition of their core similarities with their supposed enemies, so powerfully illustrated by the famous story of the cross-trench Christmas Day football matches of 1914. Yet the Subject Story did not give way after World War I. Instead, in the harrowing words of Colonel T E Lawrence, better known as Lawrence of Arabia:

When we achieved and the new world dawned, the old men came out again and took our victory to remake in the likeness of the former world they knew. Youth could win, but had not learned to keep: and was pitiably weak against age. We stammered that we had worked for a new heaven and a new earth, and they thanked us kindly and made their peace.[94]

It could have happened. President Woodrow Wilson arrived at the Paris Peace Conference with a credible articulation of the Consumer Story in his famous Fourteen Points, most of which centred on the single demand for self-determination on the part of all nations that wanted it, regardless of size. Wilson's worldview was in sympathy with the one expressed by Lal Bal Pal: rooted in a belief in the capacity and capability of everyone as opposed to just the traditional centres of power. But the old colonial powers did their deals behind the scenes, and America, for all its noble declarations, was not prepared to get too involved. Lawrence and others may have held dreams of a new order, but they were comprehensively squashed.

Instead, in Britain and the other colonial nations Subject

logic returned – and even, in the roaring 20s, briefly seemed to take on a new lease of life. In Germany, it took hold in one of its darkest ever forms, in a nation made susceptible by punishing economic oppression.

Meanwhile, in America, the Consumer Story blossomed, becoming recognisable as the story that still dominates today. When World War II broke out, the determination to reshape the world gained sufficient strength to break the shackles of the Subject Story once and for all; this time, the task of remaking the world was taken on fully.

HOW TO CHANGE A STORY

The fall of the Subject Story and the rise of the Consumer are proof that change at the level of a deep story is possible. The Citizen Story can replace the Consumer, as the Consumer replaced the Subject. This outcome, however, is by no means inevitable. The long and drawn-out demise of the Subject Story lasted over half a century and two world wars. The Consumer Story has its own old men – referring back to Lawrence of Arabia's lament – who will keep trying to solve the problems of our time with the same thinking that created them. There is also a danger of a return to Subject logic, as we're seeing with the resurgent allure of the strongman, the rise of leaders like Trump in the US, Narendra Modi in India and Brazil's Jair Bolsonaro. As the challenges we face intensify, the Subject Story's promise of protection, however empty, becomes ever more attractive.

But this time, there are good reasons why we stand a better chance. In the Consumer Story, power is more widely distributed than in the Subject (not that that's saying a huge amount), with many nations, organisations and institutions having real and significant voices and reach, not just a few Kings, Prime Ministers, and Foreign Secretaries. This means there are more intervention points available, where power can be shifted. We have the potential of the internet, despite its simultaneous perils. And now for the first time we can at least glimpse the stories that shape and drive our behaviour and, critically, our institutions.

Even so, there are pitfalls we must avoid. For one, there is the danger of blaming the individual not the system. Because of the Consumer Story's emphasis on individual responsibility, this is a common trap. If we don't see that the decisions of our politicians, for example, are in large part only the inevitable expression of the story, the risk is we think we can solve it just by voting in a new leader or party. The false solution of replacing individuals distracts us from building the mandate for deeper change; it erodes our agency.

We also have to be wary of allowing the many symptoms of collapse – which look like distinct crises – to absorb all of our attention and energy. It is far more effective and efficient to focus on the single story that gives rise to these many crises, analogous to preventing pollution upstream rather than reacting to all the disparate issues that arise downstream: the many impacts on the health of humans and animals, and on the quality of the soil and the water.

Then there is the danger of focusing on what we are against to the exclusion of building what we are for. For decades, efforts to drive change have taken the world-as-is and its entrenched institutions head on: broadly we call it the Resistance. In light of the persistence of climate crises, war, white supremacy, the patriarchy, and poverty, to name our most tenacious challenges, we might ask ourselves if our Resistance has, indeed, been futile. The contemporary Nigerian philosopher Bayo Akomolafe is among those who suggest this is the case: it is in our seeking to escape the prison, he suggests, that the prison walls gain their form.[95] If all we do is react to the dominant story, we reinforce its dominance. Perhaps the strategy itself – collision-course, expose-and-attack, hit-them-where-it-hurts – has been tainted by the old stories, accepting the terms of engagement they offer.

What if an alternative path to change opens up when we return to the concept of revelation: the moment of the veil dropping, the perspective shifting, an entirely new reality becoming apparent? The word revelation derives in part from the old French *avaler*: to sink, to lower, to get down. Instead of confronting the world-as-is head on, what if we need to get down underneath the surface? To get right to the foundations, to expose the roots of the story out of which everything is formed?

The pioneer of systems thinking, Donella Meadows, offers this definition of the work in her essay *Leverage Points: Places to Intervene in a System*:

So how do you change paradigms? In a nutshell, you keep pointing at the anomalies and failures in the old paradigm, you keep coming yourself, loudly and with assurance from the new one, and you insert people with the new paradigm in places of public visibility and power. You don't waste time with reactionaries; rather you work with active change agents and with the vast middle ground of people who are open-minded.[96]

As the Consumer Story collapses, we need to make sure as many people as possible can see that that is the dynamic at play, focusing attention on the cause not the symptoms. We must neither accept what we are given as the only possibility, as Subjects do; nor throw our toys from the pram when we do not like what is on offer, as Consumers do. As Citizens, we must propose, not just reject. We must start from where we are, accept responsibility, and create meaningful opportunities for each other to contribute as we do so. We must step up, and step in.

The Turkish novelist, activist and political commentator Ece Temelkuran offers a powerful metaphor for the work of transitioning to the Citizen Story in this time. In her 2021 book *Together: 10 Choices for a Better Now*, the eighth choice is of 'the reef over the wreck.' She describes how in 2016, a decommissioned aeroplane was deposited at the bottom of the sea off Turkey's Aegean coast. The intention was not just to create a wreck for divers. Having served out its first purpose in life, the plane would now serve a new one, becoming a skeleton structure for a new reef: 'soon the divers would be telling tales of the octopus sprawling in the cockpit or the sea turtles making love in the business-class toilet. Eventually life would transform the skeleton, leaving no trace of the wreck. The schooling and shoaling fish that had been homeless would have a new shelter. This,' writes Temelkuran, 'is what politics might look like in the coming decades.'[97] Not just politics, I would add: the wreck-turned-reef is a vision for the transformation of every aspect of our society and culture.

CROSSING THE THRESHOLD

In 1978, the historian and religious scholar Thomas Berry remarked that we were 'between stories':

The Old Story – the account of how the world came to be and how we fit into it – is not functioning properly, and we have not learned the New Story. The Old Story sustained us for a long time. It shaped our emotional attitudes, provided us with life purpose, energised action. It consecrated suffering, integrated knowledge, guided education. We awoke in the morning and knew where we were. We could identify crime and punish criminals. Everything was taken care of because the story was there. It did not make men good; it did not take away the pains and stupidities of life, or make for unfailing warmth in human association. But it did provide a context in which life could function in a meaningful manner.[98]

Berry's prescient call was for a shift to a new framework, a new deep story, that would provide the answer to the question of how we should live. He was among the first to see and state the need: the new story must be clear and compelling, its benefits widely apparent. To paraphrase Thomas Kuhn's classic analysis in *The Structure of Scientific Revolutions*, we can't have paradigm shift without a paradigm to shift to.[99] But Berry also named the challenge of the liminal period while we are between stories. The particular and very personal, day-to-day dynamics of this threshold state cannot be overlooked if we are to succeed in stepping into the new story.

The process of crossing the threshold is demanding for each and every one of us. When the cracks appear in a long-held belief, it causes anxiety and pain. As the certain world is replaced by great uncertainty, the risk is that we cling to what we know more than ever. The gravitational pull of the familiar exerts itself, no matter how dysfunctional we know the familiar to be. When we recognise this, we can hold the space for this collapse and this transition more gently, more respectfully, with greater care. Bayo Akomolafe speaks evocatively of the importance of sanctuary, a word we conventionally associate with retreat or disengagement, within the process of what he terms 'shape-shifting': '... sanctuary, in the way I think and talk about it, is this place where we gain different shapes, where we lose shape, where we compost. And how do we do that? We do it by listening, we do it by working together in ways that are probably fugitive and outside of the normative ways of producing food or money or stories. We do it by listening to our wounds, by sharing wounds, by sharing painful feelings, by sharing our jealousy and our grief. We do it by sitting with the trouble of being alive.'[100]

We can see in our world today why this transitional caretaking is so important, as the consequences of our failure to attend to it play out. When we are confronted by new phenomena that destabilise our way of life, anxiety can flip into anger. This is a particular danger with the Consumer Story, as Ece Temelkuran observes with another of her 10 Choices, urging us to 'choose attention over anger.' 'Among the emotions, anger is the most engaging and by far the most profitable.... However, this constant expression of anger, with its illusion of political or social engagement, actually makes us even more submissive.'[101] Anger pulls us back into the old story, drawing us into fighting on its terms. A key part of the defence system of the Consumer Story is that anger plays right into its business model.

Mistrust is another common result of the deeply unsettling nature of these foundational shifts at the level of story. If those in positions of power act as if there is nothing wrong, nothing to see here, our mistrust in them deepens still further. It is at these moments that many will grasp for that bizarre form of explanation known as conspiracy theories, which tend to have caricature villains at their core to blame for everything. As I write, as much as half the US population believes in the QAnon conspiracy[102]: that a Satan-worshipping, child-molesting cabal of liberal elites is in control of the government and the media, who are simultaneously responsible for Covid.

The result risks becoming a vicious cycle: as the challenges of our time intensify, we trust our leaders less, the outlets we seek in our dissatisfaction become more extreme, and our leaders in turn trust us less; seeing conspiracy theories as evidence of the madness of crowds. They become yet more inclined to stick to what they know – the old stories – denying us agency in those challenges as they engage in futile attempts to solve them for us.

We humans are uncomfortable with ambiguity, especially when

everything feels so out of control already. But there is a key distinction that can help us with the process of change. In an academic paper entitled *Towards Positions of Safe Uncertainty*, psychologist Dr Barry Mason distinguishes between 'certainty' and 'safety.' People seeking therapy generally think what they want is safe certainty: total solutions to their problems. Yet, as Mason reminds us, the next problem is only ever just around the corner: 'solutions are only dilemmas that are less of a dilemma than the ones we had.'[103] Instead, while it is essential to cultivate safety, certainty is actually something that should be actively avoided. Certainty is actually a rather rigid state in which creativity and agency cannot exist. 'If we can become less certain, we are more likely to become receptive to other possibilities, other meanings,' Mason writes.[104] The key is to cultivate safe uncertainty. This mindset is consistent with the idea of things being in a constant state of evolution and flux (leveraging neuroplasticity, in neuroscience's terms). A greater sense of agency and creativity is the result.

For those who would lead the way into the Citizen Story, the lesson is that we need to provide safety and sanctuary; but we don't need to accept the false certainties offered by the old stories. To embrace a new story, safe uncertainty is the mindset we need to cultivate not just as individuals, but as organisations, institutions, and indeed whole societies. This demands honesty, acknowledgement of the challenges, and a commitment to facing them together.

With our minds opened to the possibility of other truths, other stories, we can allow the Consumer Story to unravel. We can acknowledge its unsustainability not just from the perspective of natural resources and climate impacts, but also in terms of our record levels of depression, loneliness, suicide; our distrust in media, government, each other. It is the Consumer Story that is broken, not humanity. We no longer need to cling to it; nor is our only other option to give way to the alternative but equally false certainty of the Subject Story.

Then we can experience the Citizen revelation. We can look around and allow ourselves to see that Immy, Kennedy, Bianca, Reen and Billy don't need to be Great Men or Women. They don't need to play the role of King Sargon or Queen Victoria in the Subject Story, or of Branson, Jobs, and Bezos in the Consumer. We don't have to look up to them. Instead, they are our peers and fellow Citizens, expressions of an emerging story we can all step into and shape together. We can look back and see the times and places of Citizen emergence not as curiosities, but as precursors to this moment, when we are better equipped than ever before to embrace a bigger idea of ourselves. We can see all these, and our own time, not as causes for fear, but as sources of hope.

'There's a crack, a crack in everything,' sang Leonard Cohen. 'That's where the light gets in.' The Consumer Story has cracked, and is collapsing. That's a good thing. Now we need to get to work, showing up as Citizens in the places where we live and, crucially, in our places of work.

PART III:
UNLEASHING OUR POWER

The Consumer Story as we know it is breaking apart, and so it is inevitable that the institutions, structures and systems that were created from it, to manifest and sustain it – from the Bretton Woods institutions to the Universal Declaration of Human Rights, from representative democracy to shareholder capitalism – are no longer fit for purpose. In such times, people naturally lose faith in institutions. But new alternatives are taking shape. This time of crisis is inevitable and essential; but what happens next is up to all of us.

Much of the vital work is already under way. Immys and Biancas and Kennedys are getting organised across the world, working from outside the corridors of power to create new organisations and movements in all sectors. But the lesson of the stuttering shift from the Subject Story to the Consumer is that this truth alone will not be enough to effect the transition. In addition, we need to transform existing organisations, to open them up, invite in and thereby truly unleash the rising power of Citizens. Businesses, governments and their many agencies: these are the most potent storytellers of our society, and right now they are singing us a lullaby. Almost all of them are trapped inside the Consumer Story; until and unless we change the story from which they operate, they will serve to keep us trapped too.

This work of transformation is not easy. We have to get beyond the usual suspects of mission statements, supply chains, diversity initiatives. We have to go beneath the surface, and rebuild from the inside out. We have to get into the detail. This is because, even having recognised the Citizen in ourselves and others as individuals, the gravitational pull of business-as-usual in entrenched institutions can still drag us back to the old ways of doing things. When it comes to taking action or facing challenges, we need mental models, structures and processes to help us. If the only apparent and accessible ones have arisen from the Consumer Story, we keep using them. As we do so, our grip on the Citizen Story inevitably loosens and soon releases.

In the first years of the New Citizenship Project, I saw this consistently played out.

During a project, it would seem the veil had fallen from the eyes of everyone involved; it felt like things could never go back to the way they had been. But all too soon, once the project ended, that was exactly what happened: a reversion to the Consumer way of operating. Past clients seemed almost to feel guilty when I met up with them, as if they felt they had let me down – or maybe, that they had let themselves down.

This situation began to change when Reen and I developed NCP's model of the Three Principles of Participatory Organisations. This is a deliberate hack on what is sometimes known as 'the Marketing Mix,' a framework which identifies the 'Four Ps' essential to any business as Product, Price, Placement and Promotion. The Marketing Mix is a fixture in MBA programmes across the world, and a key part of the cognitive infrastructure of today's business world; the questions it poses (What is the organisation selling? What does it cost? Where do people find it? How does the organisation get their attention?) are exactly the questions that kept inexorably drawing organisations back into the ways of working that they were desperate to leave behind.

The New Citizenship Project response posits Three Citizen Ps: Purpose, Platform, and Prototype. Again there is a corresponding question to each. For Purpose: what is the organisation trying to do in the world? What can people not just buy from the organisation as Consumers, but instead what work in the world can we buy into as Citizens? For Platform: what opportunities does the organisation create for participation in that work? How does the organisation make involvement not just possible, but desirable, even joyful? And for Prototype: What is the starting point, the next right step, rather than trying to flip the whole organisation overnight? How does the organisation build the energy for this new way of working? How can the organisation iterate, continually revisit and improve with Citizen input, rather than seeing its role as to deliver perfection for Consumers?

In this final section of the book, I introduce these three principles, looking at each through the lens of a different sector. This is not because each is only relevant to one: every organisation in every sector needs to embrace all three if it is truly to step into the Citizen Story. However, it is true that each one of the Three Ps has landed particularly powerfully in a particular sector. Non-governmental organisations (NGOs) tend to benefit most from the challenge of Purpose, having often been overwhelmed by a corporate inferiority complex in the Consumer Story. Businesses gain most from the idea of becoming Platforms, involving people rather than just selling to them. Prototyping, the invocation to think iteratively, to go where the energy is and just begin, has proven most impactful and liberating for governments.

Of course, even when an entrenched institution – even a good number of them – does manage to transform and embody the Citizen Story, this will not be sufficient to create a Citizen Society. I have ended each chapter with a provocation and brainstorm for a truly sizeable transformative project – the Universal Declaration of Human Rights, in the NGO chapter, Facebook, in the business chapter, and Britain itself, in the chapter on government – that would amplify and scale the new story.

This part of the book is not an exhaustive roadmap for the transition to the Citizen Story. There are other tasks to take on. Vital sectors and arenas are not covered, such as finance and land ownership. The Three Ps will not be the only mental model we need, just as the Marketing Mix and its Four Ps represents just one building block of the Consumer Story.

However, as my mentor Dr Orit Gal often likes to say when she speaks of social acupuncture as a metaphor for systems change: there will always be more places to intervene than you can possibly imagine, let alone list. What matters is identifying a few good places to start.

5. CITIZEN NGOs

When I left the advertising industry, the first haven was the National Trust, the charitable organisation that works across England, Wales and Northern Ireland (with a sister organisation in Scotland) to safeguard 'buildings of historic or architectural interest and land of natural beauty forever, for everyone.' With around 350 historic houses and castles, hundreds of thousands of acres of park and woodland, and almost 800 miles of coastline in its portfolio, part of its work is the conservation and protection of these special places; part is about enabling people to experience those places. When I started, I felt a long way from home: the first week included a team day at my manager's farm, where I remember watching pigs rutting (a word new to me). I wondered what I'd done. But the three years I spent working there, and the work my New Citizenship Project team has done with the organisation since, are where many of my ideas as to how we build the Citizen Story first took shape.

The Trust was originally founded in 1895, two years before Victoria's Jubilee, in a time and place where the Subject Story remained dominant. As such, it's not surprising that – while powerful in many ways – the language of its mission carries a strong undercurrent of paternalism. In the style of Victorian-era philanthropists, Octavia Hill, one of the Trust's three founders, distinguished between the 'deserving poor' and the 'undeserving.' Her prescription for the former was getting them out of London into the countryside to see the sky and growing things – wholesome, fortifying outings that were the origins of the National Trust's approach. As for the undeserving: 'We have made many mistakes with our alms, eaten out the heart of the independent, bolstered up the drunkard in his indulgence, subsidised wages, discouraged thrift, assumed that many of the most ordinary wants of a working man's family must be met by our wretched and intermittent doles,'[105] she noted. For Hill, those who were prepared to do their bit and sought to fulfil their duty were to be given all the help in the world, including and perhaps most particularly the inspiration and benefits of beauty; those who were not and did not should be given nothing, as it would only increase their dependence. Philanthropists would fund this archetypal Subject organisation and make the decisions as to which places it should own; the poor would receive its bounty gratefully, or be damned.

After World War II, with the Consumer Story now dominant, the Trust underwent a rapid evolution. The Labour government established a National Land Fund to use money from the sale of surplus war stores to acquire property in the national interest, with the National Trust reframed as the perfect ownership vehicle. The scheme also allowed for historic houses and land to be left to the Trust in lieu of taxes; the estates of many now-struggling landowners came into the portfolio, the size of which more than tripled to nearly 400,000 acres by 1968 as a result. The organisation rapidly expanded its membership too – in 1945, a little under 8,000 wealthy philanthropists were members of what was essentially a benefactors' association; by 1968, that had grown and diversified at least a little, though with no real effort on the part of the organisation, to nearly 160,000. As the numbers grew, the motivations evolved; for many of the new breed, it was as much about a sense of shared ownership as traditional patronage.[106]

Then came the National Trust's Big Consumer Bang. A governance report in 1968 proposed extensive commercialisation. Tea rooms and souvenir shops opened across the portfolio; a Director of Public Relations was employed (Bernays would have been proud); programmes of events began, including plays and concerts. In 1984, the shift was formalised: a profit-making company – wholly owned by the charity – was set up to run the trading operations. What had been the archetypal, paternalistic Subject organisation, funded by a few patron-donors, had transformed itself into a Consumer ideal: a visitor attraction business, with not just thousands, but millions of paying Consumer-members.

Through the 2000s, the number of members swelled towards four million. But it was becoming clear that something was wrong. Membership had become a product to be bought and sold on the basis of a clear Consumer value proposition: if someone committed to an annual fee equivalent to roughly three one-day tickets, they got a free pass to the whole portfolio for the whole year. It was sold with classic Consumer single-mindedness: 'FREE entry for a year' was printed in white lettering on red backgrounds everywhere; branded binoculars and umbrellas sweetened the deal; the first year came with a three-month discount, as with car insurance. Perhaps unsurprisingly, those who found they didn't use their membership, didn't renew. Hundreds of thousands of people were joining every year, but almost as many were leaving. And there were other consequences too. Other revenue streams, like donations and legacies, were becoming harder to come by as people thought of the Trust less as a charity and more as a business. Inside the organisation, conservation staff felt less and less valued.

Meanwhile, big systemic threats to that 'forever' mission were mounting – member numbers might still (just about) be climbing, but the impacts of climate change and biodiversity loss were multiplying much faster. The 'for everyone' mission felt if anything further off: the National Trust's consumers were hardly representative; they were overwhelmingly affluent, white, conventional in every way.

When I arrived, the organisation was starting to face these challenges under the leadership of Fiona Reynolds as Director General. She would use every opportunity to talk about 'arms open conservation,' an approach that would involve and inspire visitors to make a deeper connection to the Trust's places. In service of this, she had recently pushed more power out of the head office and into the hands of those working on the ground, at the places themselves. But she knew more needed to be done, and formed a new strategy team to figure out what. I was recruited to this team.

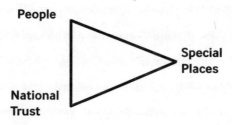

The first step was to conceptualise the task, and the relationships involved. The team produced an equilateral triangle (above), one of its sides running vertically. On the vertical side, at the bottom point, was the National Trust as an organisation, and opposite that, at the top point, were the people formerly known as consumers (I insisted on losing that language from the word go). At the third point was the object – historically significant and naturally beautiful places, 'special places' in the language of the Trust.

The diagnosis was that the organisation had in effect split itself in two: a Subject organisation, operating on the line between the Trust and special places, protecting them for people but without people (and probably, to be honest, from people); and a Consumer organisation, on the line between the Trust and people, selling the organisation to them but with little focus on the cause itself. The divide was manifest in the corporate structure (the company created in 1984 and the original charity were two essentially separate entities), and even in the office space: all the conservation staff were on the ground floor of the two-storey building, while all the marketing and visitor operation staff were on the first floor. Hardly anyone ever moved between.

The Citizen insight was that the relationship that really matters is the relationship between people and place, that final side of the triangle. And that was the relationship that the National Trust had forgotten. The team came up with a phrase to express this as the new strategic focus of the organisation: 'growing the nation's love of special places.'

Equipped with this insight, the next step was to dig back into the history of the organisation, in order to find the moments that could be 'upped' in the National Trust's storytelling, to help fix this as the focus of the organisation. We found several, but the most poignant came from the early 1930s, that previous period when, like now, the deep story was in flux. Filled with the pride and revelation of World War I, working people felt they should have a say as to the places they wanted access to, not just philanthropists – who after all were largely the same landowners who had both driven and benefited from the enclosure movement. On 24th April 1932, approximately 400 workers from the factories and mills of Manchester met at Hayfield, on the edge of what is now the Peak District National Park, and headed up towards the – privately owned – open moorland of Kinder Scout. They were met by gamekeepers and bailiffs, fights broke out, and several were arrested and sent to prison. It made national news and tapped perfectly into shifting public sentiment. What became known as the Kinder Mass Trespass played a critical part in the birth of the popular movement for national parks, which resulted in the Peak District becoming the first legally regulated National Park in Britain in 1951. The National Trust eventually took ownership of Kinder Scout in 1982, on the 50th anniversary of the Trespass. Since then Kinder Scout has stayed firmly in the spotlight of the Trust's storytelling: a mass walking event was held to celebrate the 80th anniversary in 2012; and in 2019, it was a key focus of the 'People's Landscapes' project, with musician Jarvis Cocker curating a series of events.

This triggered more change, unlocking energy and ideas that started to bubble up from everywhere as power was pushed out to the local level, where the relationships between people and place actually happened. Equipped with a small budget with which to encourage this energy to turn into meaningful action, I played the role of scout, heading out into the organisation to find and support new ways of doing things to gain momentum. A low-ranking group of employees known as 'wardens' soon put their hands up. They argued their titles and job descriptions positioned them as more like the bailiffs and gamekeepers of Kinder Scout, landowners' lackeys, than the organisers of the Trespass. They were right: they wore camouflage uniforms and their job description was essentially as outdoorsy odd job men (overwhelmingly men), who in between tasks would lurk and watch and then pop out from behind trees to protect them from children. Yet in 'growing love of place,' these people were the front line.

With help from me and my team, they conducted their own rebrand, taking the new job title of 'rangers' and making their uniforms a jolly, highly-visible red. They were trained in social media so they could engage people with their grassroots 'live on the ground' perspectives and bring the Trust alive in a new way.

So much changed. Building on the work with the rangers, my team went 'downstairs' at the head office, and turned up research conducted by the conservation teams that said that if a person develops a personal connection to nature by the age of 12, they're more likely to be pro-environment later in life. Back upstairs, this inspired a major communications focus on reconnecting children and nature, including a campaign called '50 things to do before you're 11¾' and a feature length documentary, *Project Wild Thing*, that would earn a four-star review in a national newspaper. The '50 things' – from climbing a tree to making a mud pie – were crowdsourced from the rangers in the first year, and the list then refreshed with public input in the second. 'Conservation in Action' projects started to spring up, the actual work of conserving increasingly done on show to visitors – and where possible actively involving them in doing it – rather than hidden away behind the scenes. Volunteer numbers rocketed, and the kinds of people and opportunities on offer rapidly diversified. Membership marketing changed dramatically – no more red point of sale – and the proportion of renewals rapidly went up. The commercial benefit to the organisation was pronounced: a target of five million members had been set for 2020, but that figure was achieved in 2017, three years ahead of schedule. Crucially, the diversity of the membership also increased significantly – in age, gender, and ethnicity – as did understanding and appreciation of the cause of the organisation, both among the membership and beyond.

The spirit of this work continues today, as epitomised by the leadership role currently played by the National Trust in encouraging Britain to face up to the legacy of slavery and exploitation in the days of Empire, despite strong resistance from some of its more traditionalist stakeholders. I claim no credit for this, but am cheering wildly from the sidelines. The work is not complete,

Consumer and Subject thinking never far from the surface, and the pandemic has created massive challenges, but this is an organisation proudly and increasingly doing Citizen work in the world – inspiring and enabling people to connect with and care for special places – and doing well by doing so.

UNLOCKING PURPOSE, UNLEASHING SUPPORT

Since leaving the National Trust and establishing the New Citizenship Project, NGOs – 'non-governmental organisations' – have formed the majority of our client base. This 'third sector,' as it is sometimes known, is hugely diverse. It ranges from tiny informal voluntary organisations to big charities like the National Trust and even huge international institutions like the United Nations. It includes any organisation that isn't a business (private sector) or an arm of government (public sector), and is often obscured by the prominence of these two. The range of issues involved is as varied as human activity itself, from conservation to culture to education to health and everything in between, and nothing would function without them. For all this diversity, though, the core work these organisations need to do in order to step out of the Consumer Story and into the Citizen – to experience the Citizen revelation – is remarkably similar. Almost always, it comes back to articulating the higher purpose of the organisation, releasing it from the prison of the Consumer Story. And almost always, the triangle is the key that opens the lock.

This is because the work of plotting out the triangle makes the conditioning imposed by the Consumer Story visible for what it is. The story generally manifests – as with the National Trust – in these organisations dividing themselves almost entirely into two distinct parts. First, there is the cause: the part of the organisation working directly on the object of its attention.

On top of this first function, and originally in order to 'serve' and support it, organisations have evolved a second, which develops 'business models' and 'funding propositions' to 'sell' to supporters. These are conceived of as Consumers, to be appealed to on the basis of rational self-interest, who will choose whether to support this particular organisation in competition with other organisations and causes. In the case of the National Trust, the main element of this was the highly transactional relationship with members, but the same broad mindset was also apparent in the fundraising teams, working to elicit donations from both individuals and grantmaking foundations. The money raised then funds the first part, which becomes increasingly professionalised and expert-driven in pursuit of higher quality outputs, and increasingly detached from direct relationship with people. This was where I found the conservation teams, tucked away on the ground floor of the building, when I began at the National Trust.

The result is that NGOs become moulded by the Consumer Story into heroic organisations that stand apart from and claim credit for doing things for people – and are trapped in this role because they have to keep feeding the beast, trumpeting their value in order to keep raising the money to keep delivering the outputs. They inevitably come to see their task as being to attract more support and deliver more outputs, and come to manage these strands of work increasingly separately. Fundraisers on the one hand and output deliverers on the other, they invest their energy and resources, working harder and harder in order to become ever more visibly heroic at both. And all the while, the relationships and the agency that really matters – those that are truly essential to the outcomes the organisation claims to seek – are eroding.

The task for NGOs as they step into the Citizen Story is to refocus their attention on the third side of the triangle: the relationship between their supporters and the reason for their existence. Their purpose. Instead of delivering for people and absorbing all agency into themselves as the heroes, what they need to do is build agency among both supporters and beneficiaries. In fact, they often need to dissolve this false boundary, and lose these labels altogether. This is the lesson of Kennedy and SHOFCO, and of Billy and East Marsh United: remember their insistence that the Citizens of Kibera and Grimsby alike would find and develop their own solutions, for their own communities.

Rooted in a clear sense of purpose, Citizen NGOs are able to see their task anew: to enable and facilitate, rather than sell and deliver. They become a means for many more people to organise together to deliver outcomes, rather than being self-perpetuating ends-in-themselves, organisations delivering outputs on behalf of relatively few. As a result, they both deliver far more social and environmental impact and often attract more financial support.

The change that Katharine Viner has led as editor-in-chief at the *Guardian* is another example of these dynamics in action. Wholly owned by the charitable Scott Trust, *The Guardian* can be appreciated for its role as an independent organisation within the British and indeed global media landscape, even by those who don't identify with its underlying politics or who don't love, or even read, its journalism. By the mid 2010s, with the advertising revenues that had provided the core business model for journalism increasingly disappearing into the coffers of Google and Facebook, that independent voice was facing existential threat: the Scott Trust's financial reserves were depleting rapidly year-on-year. In the course of responding from within the logic of the Consumer Story, the *Guardian* was only digging itself deeper into the hole. A big bet had been placed on Guardian Live, an initiative which would have seen the organisation take over and convert an old disused railway goods shed across the road from its headquarters in north London. Access to events and activities in this space were to become the backbone of a membership proposition rooted in exclusivity: behind-the-scenes access, preferential treatment, and so on. Not only was all this an expensive capital investment, it was also creating a beast that would demand feeding: in order to generate revenue in this way, the *Guardian*'s resources were increasingly diverted away from journalism and into the development of events and other benefits.

The Brexit vote and the election of Donald Trump to the US Presidency in 2016 stopped all this in its tracks. With almost no explicit effort on the part of the organisation, these events heralded the biggest uplifts in membership the *Guardian* had ever seen; a hastily created donations channel proved immediately popular. No paywall had been erected: people were 'paying' for the *Guardian* voluntarily. What was happening? As I argued when invited to address the board of Guardian Media Group in late 2016, people were not contributing to the organisation as Consumers in return for direct personal benefit or exclusive personal access, but as Citizens who recognised the importance of *Guardian* journalism for society and wanted to support its existence as a public good.

Throughout 2017, Viner and her team worked to formalise this understanding and in particular to articulate the purpose of the *Guardian*, in order to unlock the full potential of this instinct to support. The triangle provided the diagnosis: the *Guardian* had been on the brink of breaking itself into two organisations, with journalists understanding society on one axis, and commercial teams selling *Guardian* benefits to people on the other. By the end of the year, when she set out her 'Mission for journalism in a time of crisis' in a landmark speech and essay,[107] Viner and her team had refocused their attention on the third side of the triangle. The words of MIT journalism professor Ethan Zuckerman were at the heart of the speech and the wider strategy around it: 'If news organisations can help make people feel powerful, like they can make effective civic change,' wrote Zuckerman, 'they'll develop a strength and loyalty they've not felt in years.'[108] In seeing people as Citizens rather than Consumers, the true power and role of journalism in society became visible: it is not content for Consumers, but equipment for Citizens. We need great journalism to help us understand the world we live in, and equip us to create a better one. The donations that flowed in 2016 were evidence that people knew this instinctively, despite the conditioning of the Consumer Story. By articulating this insight more clearly, the ambition was to help the organisation work from it more of the time, more consciously, and in doing so, earn more financial support without the need to create a whole other Consumer beast to feed.

As Viner developed her thinking, the energy of the organisation shifted, the spotlight cast in new places. Filmmaker John Domokos and commentator John Harris had been making their 'Anywhere but Westminster' film series since 2009, going out across the country to speak and more importantly listen to citizens all over the country, instead of just focusing on formal politics; until this point, however, their project had existed at the edges of the *Guardian*. Now it moved to centre stage, and as it did so, began to accumulate a following of its own as well as multiple awards (including, most recently, the prestigious Orwell Prize for Journalism in 2021). Resources were reallocated away from London and Washington DC, creating greater journalistic presence out where lives are really lived. Significant investment went into recruiting from more diverse backgrounds, changing up the dynamics of the newsroom. Experiments began to explore how readers might be involved as participants in the processes of journalism, not just in the form of comments or 'user-generated content,' but in crowd-sourced editorial panels supporting and challenging journalists to get under the skin of particular places or subjects, crowdfunding experiments to support particular investigative projects, and much more besides.

And of course, the membership model changed completely. Exclusivity-based Consumer membership (whether Guardian Live or even a basic content paywall) was taken off the agenda, and the priority became delivering the public good Citizens needed, and asking for direct financial support in return. Shortly before

Viner delivered her speech at the end of 2017, direct revenue from readers exceeded advertising revenue for the first time; and by 2019, Viner was able to report that the *Guardian* was breaking even financially, with the support of over 1 million readers from 180 countries. This was an historic achievement. Throughout its 200 year history, the *Guardian* had been subsidised in some way, whether by wealthy patrons or through revenue from associated, more commercial publications.

PUTTING THE ORGANISING INTO ORGANISATIONS

After a decade of working with these ideas, I can confidently say this picture will be recognisable to almost anyone working for any NGO of any significant size. I've seen it first-hand everywhere from conservation charities like the Royal Society for the Protection of Birds (RSPB) and the World Wildlife Fund (WWF); to museums and galleries like Tate, the Smithsonian, and the Wellcome Collection; to health charities like Parkinson's UK; through to unions, be they of students, teachers, or journalists; and even umbrella organisations like the National Council of Voluntary Organisations, of which 14,000 NGOs across the UK are themselves members.

If it was working with the National Trust that first formed this thinking, it has been working with Parkinson's UK that I've found most personally affecting. In organisations like this, allowing the frame of organisation as hero to take hold (or servant, which can in practice be much the same) compounds the reduction in agency already inflicted by conditions like Parkinson's – a progressive neurological disease that gradually inhibits more and more of an individual's physical and mental functions. My father-in-law had Parkinson's before he passed away, shortly before NCP started working with Parkinson's UK. All he ever wanted to do was make a contribution and be part of things. For him, the very worst of the disease was having to have things done for him by others. It made him feel useless. Yet under the influence of the Consumer Story, Parkinson's UK – the very organisation that claimed to exist for the sake of people like him – had become a profession-alised organisation that hired ever more staff to fundraise in order to spend money on providing services for people with the disease. My father-in-law was in a recipient box, the rest of us in the family in the supporter box. He felt he was seen as someone who was incapable and to be pitied, and that was the last thing he wanted, and so he never really engaged.

It will always pain me that Charlie didn't experience the benefit of the changes that had already started before he died, and have continued since; and it will always be a huge source of pride to have contributed to those changes. Increasingly, this is an organisation deeply focused on building the power people living with Parkinson's have: Parkinson's UK is employing more and more people living with Parkinson's as staff; involving them more and more in decision-making and priority-setting (both in terms of research funding and in broader governance of the organisation); and pushing power out from the 'head office' in London to the local groups where the distinction between sup-porter and beneficiary is inherently less defined. NCP helped conceptualise the shift, and in particular created the concept of 'Team Parkinson's,' bringing together supporters and beneficiaries in one movement of people working together to make life with Parkinson's better and to find a cure – supported and connected, not served, by Parkinson's UK. This includes a pitch to those diagnosed with the disease that says 'when you need us, we're ready; when you're ready, we need you,' and has led to the development of communications based on the idea that 'Parkinson's Can,' not just 'Parkinson's Can't'; always the focus is on emphasising the essential agency and capacity for contribution that not even Parkinson's Disease should ever be allowed to take away from a human being.

A fascinating discovery popped up during the work with Parkinson's UK: it turned out the organisation had the Citizen approach deep in its DNA, and that the work to be done was more about uncovering and reclaiming that spirit than imposing it anew. In its original form, it had been constituted as the Parkinson's Disease Society, a membership organisation made up of people affected by Parkinson's (whether directly or indirectly) coming together to pool their ideas, energy and resources to make life with Parkinson's better, and find a cure. As I continue to work with these ideas, I'm discovering more and more examples where a similar original Citizen conception lies hidden beneath a Consumer layer. Nowhere is this more the case than with a group of organisations that desperately need to find a new voice: unions.

As descendants of the guilds – associations of craftspeople that taught and maintained standards around their craft, be that glassmaking, metalworking, textile-making, etc – the early vision for unions was bringing workers together so that they had a collective voice: a clear Citizen impulse. But as time went on, the

Consumer Story circumscribed their role. The membership dues – much like the membership fees of the National Trust – came to constitute the disengagement of members. The message is: 'once you've paid up, you leave the work to Central.' And that work is almost exclusively negotiating wages, benefits, contracts, and working conditions. Unions became managers of the transactional, and dues came to function as the purchase of an insurance policy. As a result, the right hand side of the triangle, the 'object,' has become hard to discern.

When NCP briefly worked with a teachers' union in the UK, the NASUWT, it was to help that team remember its original purpose as teachers organising together to discuss what good teaching entailed, so they could contribute to what schooling should look like. There was a clear object – good teaching – which constituted a reaction to the paternalistic state that was issuing orders to teachers, having decided what outcomes they needed from the education of young minds. But what the teachers' union became was all about due payment for due work: a bunch of employees looking out for their self-interest. The transaction. The indices. This, I've come to believe, is the lasting legacy of Margaret Thatcher's face-off with the miners; but reclaiming the underlying purpose and rebuilding from there is something that unions across all industries can take into their own hands. Unions can and must make a shift analogous to the National Trust and Parkinson's UK, reorienting themselves to enable their members to collaborate both to champion the importance and to improve the standards of the profession in question; not just sell them insurance and fight rearguard battles to defend their interests.

This conception of the potential bigger role of unions is a close analogy with some of the provocations the New Citizenship Project has offered in another sector that might at first glance seem far removed: arts and culture. In one project, NCP was invited to offer a Citizen provocation to the Wellcome Collection, the London museum wholly owned by the Wellcome Trust, a major global health foundation, whose mission is to connect science, medicine, life, and art. The language of the 'Collection' immediately fascinated the team, as it pinned the focus quite explicitly on the objects, not on the relationship between the people and the objects that the triangle highlights as crucial. I continue to believe there is a strong argument that it should instead become the Wellcome Collective, and that the museum and its objects should be significantly less focal than the community of people inspired by those physical things. If Parkinson's UK can evolve towards Team Parkinson's, why not?

A similar proposal again was the output of NCP's work with the British Broadcasting Corporation (BBC). Rather than simply paying the licence fee and leaving the creation and curation of British culture to the 'corporation' – as is implied by the paternalistic language of its founding mission 'to inform, educate, and entertain' – the BBC arguably needs to involve the British people much more actively. If National Trust curators are running 'Conservation in Action' sessions all over the country, building skills and understanding as they involve the public directly in their work, and if Parkinson's UK is involving people living with Parkinson's in priority-setting partnerships in their expert research processes, surely the BBC's producers and commissioning editors similarly need to open their doors, and invite the British people in? I'd go so far as to say the whole organisation needs a Citizen rebrand: what we need in today's world is less the BBC, the British Broadcasting Corporation, and more the MBC, the Movement for British Culture. As this great British institution becomes increasingly undermined by those who can only see it through a Consumer lens, such a confident shift would represent a powerful form of defence.

It is on the critical issue of the climate and ecological emergency, though, that Citizen thinking has most to offer, as a means to reinvigorate and reboot a set of organisations that have become trapped in a Consumer Story that is itself the fundamental cause of the issue they exist to address. This is evidenced by the successes of (at least the first phase of) Extinction Rebellion, and of the similarly distributed Fridays For Future and School Climate Strike movements. Greta Thunberg is a different kind of hero, one who insists on restating her own normality at every opportunity, rejecting the status offered her.

Annie Leonard, Executive Director of Greenpeace USA since 2014, is among those taking note and reforming the organisation she leads in response: she quickly realised that the organisation's model was 'allowing people to outsource their sense of agency; outsource their civic engagement to us. We were saying: give us 20 bucks a month to become a "member," and we'll go do something you could never do.... It was actually a disempowering way of interacting with the public.'[109] Since then, Greenpeace USA has undertaken a series of shifts that make it less of an organisation delivering for people and more of an open-source social movement, or as Leonard puts it, 'instead of us being the hero, being a hero among heroes.'[110]

NCP's work with the RSPB in the UK is pushing in exactly the same direction. This is an organisation with over one million members, with the potential for huge impact. The shift is challenging: their tagline at the outset of the work was 'giving nature a home' – the focus was on the conservation work, the implication that nature (the beneficiary) would be saved by the organisation, with people (supporters) on the sidelines. Indeed, as with the National Trust 'wardens' and the example of Patagonia in the Introduction to this book, there was a heavy and lingering sense that people were precisely what nature needed saving from: the problem, to be asked for money and otherwise kept as far away as possible. But the RSPB is leaning hard into the opportunity and the necessity to act as an enabler and multiplier of the amazing things people can do. NCP is now working in support of a fundamental shift to reposition the organisation in service of an open and dynamic movement for nature, driven by people, and seeing them instead as part of the solution, to be involved, inspired, and unleashed.

SWEATING THE SMALL STUFF

There is a clear 'business case' for organisations in this third sector to step into the Citizen Story, as the National Trust and the *Guardian* show. There are exciting and empowering opportunities for working in new ways across the sector's vast breadth. Yet, if I am honest with myself, progress is stuttering. One of the opportunities granted by the process of writing has been the space to ask myself why. If this is a way of thinking whose time has come, why is it not spreading faster? Why isn't every organisation on this journey? Even among those that are, why do many seem to take two steps forward into the Citizen Story, and one back into the Consumer?

Some of the answers correspond to my reasons for writing. This is a fundamentally different way of working, and this book is a way to make it visible and accessible to organisations who don't work with NCP. I also believe we will need several truly totemic projects which express and embody the Citizen Story in action on a global scale for this to truly take hold: each of the three Chapters in this section of the book will conclude with a thought experiment as to what one such totemic project could be.

But there is a more subtle, less glamorous challenge too, which pulls this dramatic work of stepping into a new paradigm back down to a much more humble level. The fact is that it's not just the big stuff that matters – the new projects, the staff cohorts rebranded. It's also the small things – the job titles and descriptions, the meeting etiquette, the room names, the measures of success. As I reflect, I realise that it's the organisations where attention has been paid to these small things that change has stuck. At the National Trust, for example, the shift from Wardens to Rangers and the launch of *50 Things To Do Before You're 11¾* were complemented by real attention to detail. The photography brief for the members' magazine changed to insist on showing not just dramatic shots of coastlines and castles, but people of all ages enjoying those places; the same images appeared in meeting rooms. Staff who were meeting one another for the first time were encouraged to share a story of a place that was special to them by way of an introduction, giving personal relationships to places an immediate presence. Marketing campaigns were assessed not just on the basis of visitor numbers or members recruited, but on increasing affinity to the cause of the organisation.

The reason this kind of work matters so much in this context is what I have come to think of as the gravitational pull of the Consumer as the dominant story of our society. Precisely because it is so deeply ingrained, it manifests in all these small day-to-day aspects of our working lives, conditioning us professionally in much the same way those 3,000 commercial messages a day operate on us personally, and on our collective culture. Each time it shows up, it exerts a pull. When National Trust meeting rooms are decorated with glorious but depopulated shots of beautiful landscapes, they implicitly condition those meeting in them to protect places from people; when staff interact only professionally, the normative frame is cold, rational, transactional; when the targets are all about quantity, the culture is all about immediate return on investment. The dominant story is ritualised into the context of everything we do; and unless we consciously act to break it, it drags us back in.

If we are to break the dominance of the Consumer Story as a society, we will need to do it methodically, down to the last mundane detail; there is no use pretending that this work is all creativity and invention. It is not. Especially in these times, it requires major effort from all of us, because every time we do things today the way we did them yesterday, we allow the gravity of the Consumer Story to pull us back into its orbit.

A NEW UNIVERSAL DECLARATION OF HUMAN RIGHTS

For all that attention to detail within organisations is essential, however, it is also true that time is of the essence. We need to become more conscious on a daily basis as individuals in organisations, but at the same time, we need to think bigger to be commensurate with the scale of the challenges we face as a human society. If we look up and out again at the broader context of this moment in time, at the collapse of the Consumer and the resurgence of the Subject Story, it's plain to see this kind of incremental, organisation-by-organisation change is necessary, but not sufficient. In parallel, this must be a moment for a radical renewal of the relationship between the individual and society, on a scale not seen since the forming of the great institutions that underpin our world today in the aftermath of World War II. What would that look like in practice? Institutional innovation on a scale equivalent to that heralded by the accession of the Consumer Story?

Try this for size: a Citizen-driven, crowdsourced renewal of the Universal Declaration of Human Rights, reinventing it as a global constitution for the human race.

The adoption of the Declaration by the United Nations General Assembly in December 1948 is rightly held up as one of the great moments of human progress of the 20th Century. Its 30 articles detail an individual's 'basic rights and fundamental freedoms' and affirm these as universal – holding for every human being, everywhere, at any time, regardless of race, culture, or creed. Over the seven decades since, the document has been translated into 524 languages; it has formed the basis of constitutions of newly independent nations and newly formed supranational associations; and it has acted as the foundation stone for the many achievements of human rights organisations from Amnesty International and Global Witness through to a whole series of country-specific actors.

Neither the Declaration nor the United Nations itself came out of nowhere; they were not without precedent. After World War I, and with the explicit (and doomed) intention of ensuring that would be 'the war to end all wars,' the League of Nations formed with a mission to promote and preserve world peace. The League itself had a precursor in the Inter-Parliamentary Union, an organisation formed in 1889 and still in existence today, whose members were and are the members of the various parliaments of the world. Both were recognisably children of the Subject Story: they brought together the great and the good of the world's most powerful nations, the aristocrats who ruled the world, with the intention of providing an environment in which they could assist one another to resolve their disagreements like the gentlemen they were. As the Subject Story collapsed at the end of the 19th Century, however, the Inter-Parliamentary Union and the League were doomed to failure. When they failed, a new approach was needed.

The Declaration responded from within the logic of the Consumer Story, and was conceived in the epicentre of that story, the United States. In his 1941 State of the Union Address, President Roosevelt set out the framework of the Four Freedoms: freedom of speech, freedom of religion, freedom from fear, freedom from want. In doing so, he was speaking a new language:

the language not of the aristocracy, but of the individual. These freedoms would be officially adopted as the war aims of the Allies, formally known as the United Nations, and formed the basis of the United Nations Charter, and then of its upgrade in the form of the Declaration. With a drafting committee chaired by Roosevelt's wife Eleanor, 'First Lady of the World,' the document was written in language that could and indeed should be understood by anyone and everyone. The United States would lead a world where the individual was king and must understand his rights, where the great and good stood to serve, not rule; as a result, world peace would be secured, and all humanity enabled to flourish.

Except, of course, peace was not secured, and – despite notable advances – all humanity has not flourished. The world has too often turned a blind eye to oppression, in the name of trade, economic growth, and national self interest. Today, as the Consumer Story that birthed them collapses, the flaws of both the Universal Declaration of Human Rights and of the United Nations are becoming evident, just as with the League of Nations during the unravelling of the Subject Story. The League depended on the world's aristocrats to be confident and outward-looking at a time when they were becoming increasingly insecure, defensive, and obsessed with their own immediate self interest; the Declaration depends today on individuals in exactly the same situation. In this harsh light, the document is exposed for its form as a declaration rather than a legal treaty; it lacks binding powers to force action, and seems now – as do many of the idealistic initiatives of the Consumer era – an object more performative than substantive, more for show than for impact. Developed and defined for the world by the few self-appointed 'champions of peace,' then served up for adoption by the many, its influence is further undermined by the hypocrisy of those few: the failure of the United States to honour the rights of the black community; the reluctance of the British to deliver full rights to their former colonial subjects, and now the desire to withdraw from the jurisdiction of the European Court of Human Rights. Perhaps most fundamentally, with its extreme focus on the individual, the Universal Declaration of Human Rights ignores the fact that the greatest abuses are not of individuals, but of religious or racial groups – communities and collectives – at the hands of their own governments. At exactly the moment the Universal Declaration of Human Rights is most needed, it seems to have least to offer.

Could we apply the logic and tools of the Citizen Story to reinvent the Universal Declaration of Human Rights, making it live up to its potential?

This time, it would need to be composed not just by a drafting committee, but by the Citizens of the world. It would be run as an 'outsider' process, led and facilitated by a coalition of international and national NGOs from across the world stepping into the Citizen Story together. Going back to the concept of purpose and to the triangle again, organisations like Amnesty and Oxfam would need to transcend a status quo in which they are too often stuck competing with each other to raise money on the one hand, and spending it heroically on the other. Instead, recognising that their higher purpose is in fact something they share, they would come together as collaborators in a project that would redefine their foundations. They would facilitate an enormous collaborative process, giving people across the world – 'developed' and 'developing,' 'supporters' and 'beneficiaries' – the opportunity and the space to work together, creating genuine agency in contributing to and shaping the output.

With the developments in participatory techniques of the last decade, such a process is possible to imagine, putting together elements that have already been proven to work all over the world. The existing Declaration could be seen as version 1.0; it could form the basis of the process, to be hacked and edited. An open, distributed, global conversation could see proposals for revisions and new articles encouraged from quite literally everywhere – exactly as per the process that developed Mexico City's crowd-sourced constitution.[111] These should then be deliberated on and prioritised by a randomly selected, representative sample of the population of the world, coming together in the form of a Global Citizens' Assembly – exactly as per the Global Citizens' Assembly that produced the People's Declaration for the Sustainable Future of Planet Earth in the run up to the Glasgow climate talks in November 2021.[112] This Assembly phase would take place over the space of a few months, and during this time, the NGOs involved would seed and support distributed 'kitchen table conversations' around the world, inviting and inspiring people of all backgrounds to follow and feed into the discussion.

The output of this process would be a crowdsourced constitution for humanity, in draft form: a statement of the conditions that we collectively believe need to be in place for human flourishing. With such widespread input, powered by collaboration across so many organisations, this statement would have unrivalled legitimacy as an expression of global will.

The organisations involved in its development would then be in a position to challenge the governments of the world to adopt it formally, or explain why not. This task could be separated out into layers: the international NGOs might focus on the United Nations; national NGOs on their respective governments. Once adopted, it would form a legal basis that the existing Universal Declaration of Human Rights has never had. We could do this in time for its 75th anniversary in December 2023.

A pipedream? Perhaps.

But it would be better to reinvent before the next global war than in its aftermath.

6. CITIZEN BUSINESS

The little town of Stonehaven, just south of Aberdeen on the east coast of Scotland, is an unlikely birthplace for a radical experiment in a new way of doing business. It is a pretty but sleepy little harbour town with a population of around 10,000, more on the relatively rare dry days of summer, fewer in the long dark winters. But when James Watt grabbed a Sierra Nevada Pale Ale to wash down his fish and chips one evening in the early 2000s, the die was cast. Watt's previous experience of beer, he claims, had been a single tasteless lager on holiday with friends; he'd since chosen the Scotsman's option of whisky at every available opportunity. But in that moment, sat on Stonehaven's harbour wall, he fell in love with craft beer, and he wanted literally everyone else to experience the same.

Watt and his friend Martin Dickie began their own homebrew experiments, initially trying to replicate the Sierra Nevada, and then branching out from there. The breakthrough came in 2007, when the two 25-year-old brewers entered a competition run by Tesco, then Britain's largest supermarket chain. Their beers came first, second, third, and fourth. The prize was a contract to supply Tesco with more than twice as much of their winning beer, Punk IPA, as they were at that point capable of producing. BrewDog – named for Dickie's chocolate labrador puppy, Bracken – had arrived.[113] Fifteen years on, the company employs over 2,000 people, operates 100 bars in countries around the world, and has featured on the *Sunday Times* list of 100 fastest growing UK companies almost every year since 2008.[114]

The pair were hugely ambitious right from the beginning, and in those early days that ambition was about more than just brewing and selling beer, growing the company, making their fortune. It was about breaking the stronghold of the big industrial brewers, the companies whose efficiency- and profit-driven dominance had reduced the beer industry to the lowest-common-denominator, tasteless disappointment that had led Watt to reject the drink outright. They didn't just want to make a different kind of beer. They wanted to run a different kind of business.

At its heart would be a clear and simple expression of their driving purpose, capturing the spirit of that moment on Stonehaven's harbour wall: what BrewDog would be about was not just selling beer, but 'making everyone as passionate about craft beer as we are.' That focus led them from the beginning to see their own products as part of something bigger; even from the early days the pair lent and even gave some of their profits to others who wanted to start and expand their own independent craft breweries, seeing these more as allies and fellow travellers to be supported than competitors to be crushed. Fellow Citizens in a new business ecosystem, we could say.

The most fundamental difference that clarity of purpose made, though, was that Watt and Dickie followed it through to the logical conclusion of seeing their customers differently. In the original BrewDog frame, people were not just consumers, transactional sources of money who choose this brand or that. They were fellow beer lovers, existing and potential. Peers with a shared passion, even friends, not faceless masses to be represented only by numbers on the management accounts. They were part of the cause, indeed participants in the cause, to be involved and inspired, not just sold to.

This mindset led to BrewDog doing a whole series of things that wouldn't occur to most other businesses. Watt and Dickie would sell beer-making kits as well as beer, and even open source all their recipes.[115] Most companies would see giving away key intellectual property as foolhardy. For BrewDog's founders, it was a no-brainer: they just wanted more people to love craft beer as much as they did, and of course that meant being able to brew their own. They invested a significant amount of money in the spread of a previously niche qualification system for 'cicerones,' the beer equivalent of wine sommeliers, because they wanted to raise the level and seed a whole culture of beer appreciation.[116] And most significantly, they took every opportunity to make their customers their fellow business owners, creating a community of 'Equity Punks' that at the time of writing numbers a little over 175,000, and has invested somewhere in the region of £100 million into the business.[117]

It is hard to emphasise just how radical this step was when Watt and Dickie first started down the path in 2009. Today a whole series of intermediaries exist to enable what is now called equity crowdfunding, including CrowdCube, Seedrs, and more; back then, the first seven legal advisors Watt and Dickie approached told them what they were trying to do was impossible. Setting up the first round of Equity Punks investment cost the business £100,000 at a time when there was only £40,000 in the bank.[118]

What this community has meant to the business goes beyond conventional shareholder investment – where shareholders-as-Consumers put money in on the understanding that more money will come back out, as quickly as possible – and the relationship created was and indeed still is very different. Thousands of Punks gather for the AGM, Annual General Meeting in Consumer-speak, but which in BrewDog world stands for Annual General Mayhem. These gatherings involve the Punks in the governance of the business, but they are also festivals of shared purpose, with live bands and free-flowing beer (both BrewDog's own and that of the other breweries they have supported). Punks have been involved in the company's advertising, and sometimes have even been the advertising: the 2017 I Am Punk campaign saw images and stories of shareholders posted up in all sorts of locations.[119] They participate in product development, through regular tastings and naming competitions. Those who want to get more involved can join the Beatnik Brewing Collective, a subset cooperative which develops new recipes for the wider company. This relationship with its customers has led Brewdog to some important decisions: Watt and Dickie openly credit the influence of the Equity Punks on the decisions to become a living wage employer, to build pioneering eco-breweries, and to launch an impressive corporate environmental agenda under the banner of Brewdog Tomorrow.

BEYOND PURPOSE: BUSINESS AS PLATFORM

The move from profit to purpose as the primary focus of corporate decision-making is an important part of the move from Consumer to Citizen as the story of society. When Milton Friedman declared in his 1962 book *Capitalism and Freedom* that 'there is one and only one social responsibility of business – to use its resources and engage in activities designed to increase its profits,'[120] he was codifying pure Consumer logic. The explicit argument was that businesses should look out for their own narrow self interest as corporate entities, paying their employees as little as possible, taking on responsibility for as few social or environmental impacts as they could get away with, and serving the narrow financial interests of their shareholders as fully as they could, because the aggregation of that self interest was how the best outcomes for society would result. If each business maximised its profits, people-as-Consumers would benefit from better products and services, government would be freed up to focus on what only it could do, and all of society would benefit. All else was politics: it was for government to set the parameters, not business.

Over the last decade, prominent academics at established institutions have systematically dismantled this argument. Instead, it is now increasingly accepted that creating 'shared value' across all stakeholder groups (employees, communities, customers and shareholders) is the appropriate goal rather than narrow shareholder value, and that the best and indeed most profitable businesses achieve success by aiming to 'grow the pie' for all of society, not just take the biggest possible slice. Putting purpose first, and seeing profit as a by-product, is the increasingly universal prescription. Following this theory, new business networks and even new legal structures have emerged and spread, most notably the B Corporation movement and the associated US legal structure of the Benefit Corporation. Both these see businesses retain their commitment to profit-making, but explicitly put that in service of social and/or environmental purpose. Certified B Corporations – of whom there are now over 3,500 across 70 countries[121] – have to achieve a minimum score on a rigorous assessment across a broad range of considerations, from governance and employee relations to community and environmental policies and commitments. Company directors also have to sign up to a 'Declaration of Interdependence,' affirming their belief 'that we must be the change we seek in the world; that all business ought to be conducted as if people and place mattered; that through their products, practices and profits, businesses should aspire to do no harm and benefit all; and that to do so requires that we act with the understanding that we are dependent upon one another and thus responsible for each other and future generations.' I am proud that the New Citizenship Project was a founding member of the UK contingent.

But while all this is both necessary and important, there are two reasons why the shift from profit to purpose is not my focus here. First, it's in train: once the *Harvard Business Review* and McKinsey and EY and the rest are talking about corporate purpose, the tone is set. No need to add my voice to the clamour. Indeed, in my experience with NCP, the challenge of articulating purpose is actually much more profound for the organisations we met in the previous chapter: they seem to take purpose for granted because they are not formally profit-oriented. The private sector is much more on the case.

The second reason is more important. It's simply not enough. Many of the most lauded 'purposeful' corporations are actually doing most to perpetuate the Consumer Story. To explain this, I return to the days when I was still working in an advertising agency, and to what I thought at the time was going to be a dream project. The client was Unilever, the food and healthcare multinational which was then and remains to this day a paragon business of 'sustainable consumption.' Its then CEO Paul Polman remains feted as one of the great champions of purposeful business, in particular around the issues of climate change and environmental degradation. Having played a major part in convening the Round-table on Sustainable Palm Oil (RSPO), work that was critical to reducing rainforest clearance across the world, Unilever's marketing team came to the agency I was working at with a brief to promote sustainable palm oil, and invite its customers to be part of the solution.

I knew the background: Unilever's part in RSPO had been initiated in response to direct action from Greenpeace, which saw activists in orangutan costumes scale several of the company's buildings and drape banners which read 'Stop destroying rainforests for palm oil.' I proposed that we make recognition of this dynamic the core proposition of the advertising. The concepts featured images of those protests, with the line, in the voice of Unilever: 'This was our wake up call. This is yours.' The idea was to celebrate the role of activism and its potentially powerful positive relationship with business, encouraging 'consumers' to recognise this at the same time as encouraging them to seek out the ethical alternative.

In testing, the ads were hugely impactful. People wanted to know more – about palm oil, but also about Unilever and indeed Greenpeace. The ads made them think differently about all three: making palm oil a major issue, repositioning Unilever as a committed ethical business, and framing Greenpeace and activists in general to be embraced as courageous rather than feared as dangerous. The ads went up through the chain of command at Unilever for approval.

They never came back. The budget was cut, and a replacement was dictated to us: beautiful rainforest imagery, with the line: 'What you buy in the supermarket can change the world.' The message was reduced from a powerful endorsement of the role and power of activism, to a lullaby of consumption. Instead of taking the opportunity to celebrate and invoke agency, Unilever had chosen instead to say 'Hush little people. Just go shopping. We'll fix it.'

I've seen the same dynamic over and over again since. Purpose-driven firms are radically rethinking sourcing practices, product design, value chains, and so on, and that's to be celebrated. But while they do these things for us, behind the counter, they're all too often still thinking of and communicating to people as Consumers – and contributing massively to those 3,000 messages a day, to that underlying Consumer Story. More than that, when communicated in the style of that Unilever ad, these 'purposeful' messages can be deeply destructive. They take the Citizen instincts that we all have and channel them into the act of consumption, telling us that our purchasing decisions represent not just one expression of our agency in the world, but the limit of it.

In this context, the New Citizenship Project challenge to business is that purpose is necessary, but it is not sufficient. What is also needed is the next step: to open that purpose up for employees and shareholders and customers to participate in, in meaningful, creative, joyful ways. Only purposeful businesses – businesses that exist to do something beyond simply making profit – can do this, because people will only participate in purposeful work.

This is what I mean when I talk about Platform as the second P: businesses need to become platforms enabling people to undertake purposeful action in the world, as opposed to trying to deliver on purpose for people, without our involvement. It's not about handing over the keys to the castle, becoming a user-generated company; it's about creating structured opportunities for people not just to buy products and services from the business, but to buy into what the business is trying to do in the world. It's only when this happens on a widespread basis that the story that businesses are telling will truly change. But this is hard work. In terms of language, people simply are Consumers in the parlance of the corporate world, and as a result the accompanying unspoken assumptions as to motivations and capacities are deeply embedded. This makes the possibility of anything else incredibly difficult to see, let alone step into.

THE SEVEN MODES OF EVERYDAY PARTICIPATION

This is why I immediately found BrewDog such a powerful case study. The company's purpose, 'making everyone as passionate about craft beer as we are,' may not seem the loftiest, but is authentic and embodies a respect for craft and passion. It also opens the space for participation, because it is not about doing something for people, but about inspiring something in us. At its best, what BrewDog has done has been to create opportunities for its customers not just to be Consumers of beer, but to be participants in that purpose – to become brewers in their own right with the help of the DIY Dog home-brewing kits, to become qualified cicerones, to become investors in the company and thereby the purpose as Equity Punks, and so on.

Talking about this seemed to switch a light on for NCP's corporate clients, making real for them the new relationship that becomes possible between people and business when business has a purpose, and becomes real when that business then creates joyful, meaningful opportunities for people to take action that includes but goes beyond consumption. So the team started to look around for more case studies, and to piece together a set of prompts that could make this way of thinking and acting more visible and accessible. Equipping purposeful businesses (and indeed other organisations) of all shapes and sizes to step into this, NCP's Seven Modes of Everyday Participation are each framed around what the Citizen can do. Here's a brief rundown, ordered roughly by ascending level of commitment on the part of the Citizen.

MODE 1: TELL STORIES

The prompt question here is: How might your business hand over the mic? BrewDog's I Am Punk campaign is the archetypal example, letting customers speak for the business instead of the business broadcasting to the world. The campaign began at the 2017 AGM, where 2,500 Equity Punks had their portraits professionally photographed and told their stories of why they had invested. These were then posted and shared widely across social media, celebrating the community in all its diversity, 'from Emily the pipeline engineer to Austin the wheelchair basketball enthusiast to Kay the offshore worker/pole dancer.'[122] This spread the BrewDog passion for craft beer to over 7 million people, even before a second phase saw printed posters distributed to Equity Punks and fly-posted across the UK. As per the introduction to this book, imagine the cultural impact that Patagonia might have with an analogous 'I Am Nature' campaign...

MODE 2: GATHER DATA

This mode roughly corresponds to what is sometimes called Citizen Science, whereby large numbers of people get involved in gathering data that is crucial for research purposes, flipping the usual corporate model of people being the object of the data gathering exercise: How might you tap into people as active researchers not just passive guinea pigs? One business NCP has pushed to adopt this mode is the online retailer Buy Me Once, whose purpose is all about breaking the cycle of 'designed obsolescence,' where products are effectively built to fall apart so that people will replace them. Its theory of change is essentially rooted in getting people to 'buy once' from them, something that they love and that lasts, instead of buying ten times from a standard retailer. Research into products and producers is a vital part of the Buy Me Once offer, and increasingly the company is tapping into its customers to help them do it, asking them what products they want on the site, what things have let them down, what could be made better, and what things they love.[123] This is the first step in a strategy that is positioning Buy Me Once less as a transactional retailer and more as a community. As one of the team shared with me: 'The longevity of the things we own invokes such visceral reactions, from disdain to devotion, in all of us. In that way our mission is inherently grassroots, and the power of all that emotion, if it can act as a collective, would be tremendous.... We've learnt so much, and we've found our audience very ready to be engaged in this way. It has also, as it happens, helped us sell more stuff. Which helps.'[124]

MODE 3: SHARE CONNECTIONS

It is long-accepted marketing wisdom that existing satisfied customers are the best recruiters of new, and this can be structured into a constructive, creative Citizen opportunity, asking: How might customers share our purpose, not just our products, with their friends and family? Once again BrewDog provides my favourite example for this mode, with its commitment to build a bar, each of which is effectively a temple to craft beer, wherever there are more than 200 Equity Punks.[125] As with telling stories, I'd love to see this adopted by the kind of environmental and ethical businesses that generally tend to default more to guilt than joy in their motivations. A recent advertising campaign for oat milk brand Oatly, for example, basically encouraged people to shame their family members for consuming dairy products. What if, instead, Oatly promised an oat-powered party in every area where it sold a certain amount of product?

MODE 4: CONTRIBUTE IDEAS

This is one that's really starting to gain momentum, with lots of businesses starting to tap into the ideas of their customers to drive them along, asking: How might you ask for your customers' experience and ideas, not just their money? My favourite example is the start-up UK brand Tortoise, co-founded by former BBC Director of News James Harding and former *Wall Street Journal* President Katie Vanneck-Smith. As part of its purpose in slowing news down (hence the name), and taking time to make sense of what's really going on in the world rather than just react to the immediate, Tortoise's editorial team hosts weekly online 'Open News Think-Ins' at which they discuss the news stories of the week with the subscriber community. They start the conversation themselves, but then open to the virtual room, actively seeking ideas from everywhere for what should be covered, and how.

MODE 5: GIVE TIME

Tortoise is only one of a number of new journalistic enterprises approaching the task of making the news in a participatory spirit. The first big mover was Dutch brand *De Correspondent*, so named because its operation is built around a series of issue-specialist 'correspondents,' each of whom sees their role as being to lead and guide an inquiry into a that issue by eliciting the help and contribution of their readers. 'We don't call them journalists, we call them conversation leaders,' says COO Ernst-Jan Pfauth, 'and we don't call them readers, we call them expert contributors.'[126] This model, for example, saw the environment correspondent work with a series of readers who were past or present employees of Royal Dutch Shell to investigate what the management of the oil giant had known about climate change when, and hold the company to account. Food retail is a space where this is long-established practice: the Park Slope Food Coop in Brooklyn, New York, requires each of its 17,000 members to contribute three hours working time a month as part of keeping its prices down, [127]and has inspired similar models in the People's Supermarket in London's Bloomsbury and HISBE (How It Should Be) in Brighton. UK telecommunications operator Giffgaff operates its customer service function as an entirely peer-to-peer model, its customers advising and supporting one another in return for either discounts or charitable donations from the company.[128] The point in all these examples is that the customers are the people who actually know best, since they use the product or service in question: How can you enable the people who know best to do the work with you?

MODE 6: LEARN SKILLS

Seeing people as participants in purpose rather than Consumers of products opens up the realisation that we occasionally, if not often, want to do for ourselves what a business is currently doing for us. Providing a learning experience is often lower cost, more sustainable, and creates a deeper relationship than simply selling the product. BrewDog's adoption of the cicerone qualification is a great example again. But there are whole business models here. One example is Yuup, founded by close friends of the New Citizenship Project. Its model provides the infrastructure for local independent businesses (florists, cafes, and so on) to offer learning experiences as a complementary revenue strand. [129]In doing so, Yuup allows these businesses to tap into a huge competitive advantage they have over the big chains that can easily out-compete them as pure product retailers, with huge potential to support the resuscitation of high streets across the country and beyond. Yuup is rapidly taking off in its launch city, Bristol, and is set to spread, taking the question for this mode with it: How can your business teach its customers to be participants in its purpose?

MODE 7: CROWDFUND

This final mode is the most direct expression of people buying into a company not just buying from it; while this is still essentially a transaction, the fact that the purchase is made in advance of the product being available creates space and time for relationship and community. Tech Will Save Us, another company founded by friends of NCP, has put this mode right at the heart of its strategy. Its products are DIY technology kits designed to enable children to build their own instead of buying; its higher purpose to cultivate a Citizen rather than Consumer relationship with technology, where young people can understand and deploy technology, not just use it blindly. Working with crowdfunding platform Kickstarter as its primary outlet has been a natural expression of this purpose: every product was developed and 'sold' first as a concept and only later through traditional retail channels; in doing so, the company built a direct relationship with its most engaged customers, and was able to work with them in many of the other modes above, as voices in its marketing, sources of ideas for product development, and much more.[130]

WHERE IS THE POWER?

When I talk about this concept of everyday participation, there's a challenge that often comes up. 'Don't these modes,' I'm asked (or sometimes told), 'actually just represent the Consumer Story co-opting the Citizen?' After all, businesses tend to profit by using these modes, increasing sales of their products as well as getting free input from their customers.

We've been through 'greenwashing,' whereby businesses undertake superficial initiatives to make themselves appear more environmentally friendly but ultimately little changes; how can I be sure this won't just lead to 'Citizen-washing'?

It's a challenge I take very seriously, and one which takes me back to BrewDog and some real harms the company has been guilty of in recent years. BrewDog often sailed close to the wind, with PR stunts of questionable taste playing a significant part in business strategy (for example, the presentation of its 55% alcohol-by-volume 'End Of History' beer in the stuffed dead bodies of squirrels and stoats)[131]. But around the beginning of 2017, questionable turned into outright wrong.

In March 2017, it launched a new vodka brand called Lone Wolf at around the same time as a brother-and-sister team Joshua and Sallie McFadyen opened an independent pub in Birmingham called... The Lone Wolf. BrewDog's lawyers forced the McFadyens to change the pub's name. On online forums, Equity Punks called out how inappropriate such bullying of a small independent business was in light of BrewDog's purpose and ethos. James Watt stepped in, withdrew the action, and paid all the costs incurred in full – as well as inviting the McFadyens up to Aberdeenshire, where he and Martin Dickie treated them like royalty, even making a batch of gin together. [132]

The big disappointment, though, came in April, when Watt and Dickie sold a 22% stake in the company to private equity firm TSG Consumer Partners for £213 million, valuing the company at over £1 billion.[133] £100 million went to Watt and Dickie. £100 million went into the business. Despite the huge sums involved, the opportunity for Equity Punks to sell their shares was strictly limited: they could sell only up to 15% of their total holding, or a maximum of 40 shares, whichever was smaller;[134] at the same time, it became clear that the nature of the shares gave Punks no meaningful power in this huge decision. Only the previous year, James Watt had reacted with outrage when AB InBev, the brewery giant which owns Budweiser, Stella Artois and Beck's, added Camden Town Brewery to its holding. Watt had posted a video on Twitter railing against the sale: 'BrewDog will never sell out.... Not ever. Mega corporations care about costs, market share, dividends, valuations. They don't care about beer.'[135]

At the time of writing, a new storm had just erupted. In an open letter circulated on Twitter, 61 former employees accused the company, and Watt in particular, of creating a 'culture of fear,' in which workers were bullied and 'treated like objects.' Identifying themselves collectively as Punks With Purpose, they wrote that 'growth at all costs has always been perceived as the number one focus for the company.'¹³⁶ In the same month, the winner of a recent Willy Wonka-inspired marketing campaign to win a 'solid gold' beer can found out it was merely gold-plated.

What are we to make of all this? Certainly, BrewDog has not lived up to all of its ideals. Having watched Watt and Dickie closely, I believe their original motivations were authentic, but that they have latterly fallen foul of 'Founders' Syndrome' – where one or more founders maintain disproportionate power and influence following the effective initial establishment of the organisation, leading to a wide range of problems. Watt in particular has come to believe his own hype, forgetting that it is the Equity Punks, not just him, who have made the company what it is. But without the voice of the Equity Punks structured explicitly into the company, that voice has become lost.

That said, the BrewDog story is not over – and signs are that the Punks are starting to grasp the influence they do most certainly have, even if it is not formal power. The energy is building again, on the forums and on social media, and it's getting noticed. A recent *Financial Times* article carried the headline 'Punk Rebellion: BrewDog's crowd-funding investors start to lose faith,' observing that there is a very real reason why the company needs to respond to Equity Punk concerns: 'BrewDog's latest crowdfunding round values it at about 10 times net revenues... remarkable considering that listed brewers such as Heineken and Molson Coors trade at enterprise values of two to four times listed revenues.'¹³⁷ BrewDog is worth what it is to investors, because of the Equity Punks. If Punks depart, so will the investors.

BrewDog is flawed, but it has pioneered an approach that other businesses can learn from, and go beyond. The real opportunity is to embrace modes of everyday participation, but critically also to structure meaningful Citizen power into the governance of the organisation. BrewDog might, for example, give Equity Punks representation on the company's board. Going a step further, NCP is working with two different businesses at present to explore how citizens' assembly-style approaches might be brought into their governance. There are some inspirational case studies out there too: renewable energy supplier Good Energy has created a Good Future Board, for example. This body, which has genuine power in the business, is entirely made up of school age children, on the basis that 'the time is up for hearing from grown ups on climate change.'[138]

The greatest opportunity, however, lies with a subset of businesses that have this kind of meaningful Citizen power right at their core, but who – locked in the prison of the Consumer Story – have not yet stepped into the opportunity of everyday participation. This is the cooperative movement, which I want to focus on in some depth. NCP has been working with the Co-operative Group, or Co-op for short, for several years.

Created while Britain was still deep in the Subject Story, the Co-op was very much a Citizen organisation at the outset. In 1844, local tradesmen in Rochdale, just outside Manchester, pooled their resources to buy simple, safe, good quality products like flour and candles that they would not be able to afford alone, and then sell them on to an increasing number of fellow capital contributors. They called themselves the Rochdale Society of Equitable Pioneers, and from the very beginning, the governance structure they adopted gave one member one vote, regardless of the amount of capital invested. The model succeeded, scaled, and spread. Right through to the 1960s, the resulting family of cooperatives remained massively distributed, with thousands of separate local co-operative societies all over Britain and indeed the world. You had to be a member to shop at your local store, and to get the benefit of the members' dividend, which made your shopping cheaper. Membership also enabled you to vote at the Annual General Meeting, to have a say in governance. All the local stores were in turn owning members of the Cooperative Wholesale Society (CWS) – each store was a single member, holding a single vote in the governance structure – which at its peak had enough resources to buy its own ships and import food, and also to buy land and run its own farms.[139]

Until as recently as 2014, it was one of Britain's biggest farmers, owning 16,000 hectares of land across the country.[140] The CWS even established a bank, again with its own cooperative governance structure, and started offering financial services direct to the public, with the money invested into the local cooperatives, as well as the wholesale ventures. It was a whole, thriving, alternative business ecosystem.

As the Consumer Story took hold, though, and challenges to market share arose from the fast-growing supermarket chains, the model began to creak, and ultimately power flooded in from the local to the central. In the 1960s, changes to retail pricing regulation effectively removed the price advantage of the dividend, allowing supermarkets to offer lower prices through promotions, undercutting and loss-leaders. With local cooperatives struggling, the CWS in effect became a corporation called the Co-op Group, taking ownership of almost all of them. Members of the local cooperatives saw their membership transferred into the big beast, which was far more distant and less relatable.

In the aftermath of the global financial crisis, the repercussions of all this somewhat panicked centralisation nearly brought the whole group tumbling down. The Co-operative Bank was at the epicentre of the problem. When power poured into what had been the CWS, all the various governance structures were simply thrown together. To become a board member of the bank, you first had to serve several years on a now entirely powerless local board; this made you eligible for election to the national board (itself too large to be particularly effective), where you had to serve several more years to become eligible for election to the board of the bank.

The result? A huge bank ended up governed by a board arguably akin to the worst kind of parish council, its members more motivated by perceived status than meaningful opportunities to contribute.

It all came to a head in the summer of 2013 and in the figure of Chairman Paul Flowers, a career local politician and former Methodist minister. A £1.5 billion 'hole' was discovered in the bank's finances.[141] Flowers resigned, and was later that year caught buying a range of illicit drugs, earning himself the nickname of 'the Crystal Methodist.'[142] The bank had to be extracted from the group and floated on the stock exchange, and now trades as the Co-operative Bank PLC – a situation that does little for public understanding of cooperativism, given that a public listed company is by definition owned by shareholders, and is therefore a fundamentally different kind of organisation from a cooperative. The wider fallout nearly took down the whole Co-op Group, ultimately forcing the sale of most of its assets (including the farms), and putting the whole organisation into survival mode until very recently. Gradually, the core business came to operate more and more like any other supermarket. Until recently, eight million people across Britain might have had membership cards but few had any sense of the original meaning – they had become just like any other supermarket's loyalty cards, a means to offer discounts to Consumers, not open up participation to Citizens.

Now though, with at least a little credit due to a pestering presence from the New Citizenship Project, things are starting to change. Working primarily with a team whose function was formerly identified as research and insights, NCP has supported them to rebrand as a Member Participation team. This team has supported experiments across the different parts of the business, inspired by the Seven Modes: in recent years, this has seen the Co-op crowdsource recipes for new food products, co-create a new travel insurance offering, invite widespread input into its new food sourcing policy, and much more. Perhaps the most significant achievement to date came during the early months of the pandemic, when it started to become clear to the Co-op's Funeralcare team how many people were being forced to grieve alone. The Member Participation team helped them work with Co-op members to establish a network of online, member-led 'grief cafes,' where members come together to share at least an online space to feel less alone with their pain and loss, with the Co-op providing the space and the checks to ensure the safety of that space. More ideas and projects are emerging all the time, the momentum is building, and the impact is growing.

Ultimately, what we need for Citizen business to take full hold is for the cooperative model to expand massively, and for the businesses in it to step into the Citizen Story. In order to recognise this for the very real possibility it is, we first have to see just how widespread co-operatives already are in our lives today. While in Britain the strength of the movement has been disguised by the travails of the Co-op Group and confused by the abomination of the Co-operative Bank not actually being a co-operative, the global numbers brook little argument. In 2020, the World Cooperative Monitor found that over 1 billion people – at least 12% of the world's population – are members of at least one co-operative; that co-operatives provide work for 10% of the employed global population; and that the world's three hundred largest co-operatives have an aggregate turnover in excess of US$2 trillion.[143] The Spanish business group Mondragon, the largest co-operative in the world, employs and is owned by over 80,000 people. Co-operation is everywhere, and it's big.

Once we understand this pre-existing scale, we start to see what might otherwise have seemed like a quirky little micro-trend as a phenomenon with the potential to change the world. The term 'platform co-operativism' was first coined by self-described schol-ar-activist Trebor Scholz, an Associate Professor at the New School in New York City, in 2014.[144] He was the first to spot a bifurcation in what was then known as the sharing economy. While the rest of us were just beginning to get our heads round the very idea of businesses like Airbnb, Uber, and the like – 2014 was also the first year Airbnb exceeded the number of rooms offered by any single hotel chain – Scholz had realised the danger that these models would simply represent a new concentration of power, and was setting out the opportunity offered by bringing co-operative ownership to bear on this 'platform' model of business. If these new companies were going to make their money not by offering goods or services themselves but by running the platforms that enable us to provide for one another, then, Scholz argued, we needed to own the platforms.

Fast forward seven years or so, and there are co-operatively owned alternatives gathering momentum against every one of the new giants. For eBay, read Fairmondo; for Airbnb, Fairbnb; for Uber, a whole array of city-based ride-sharing apps, some owned by local drivers, some by all local citizens. The list goes on. There is a very real surge building here. But as with the idea of a crowdsourced, Citizen-driven reinvention of the Universal Declaration of Human Rights in the last chapter, we need something else too. A big, totemic transformation of a big, totemic business.

CITIZEN FACEBOOK

Back in late 2015, inspired by discovering the work of Trebor Scholz at a conference earlier in the year, I had what I still think is the biggest and best idea of my life. Sitting at my desk late one evening, I read about the forthcoming launch of the Chan Zuckerberg Initiative, which would see Mark Zuckerberg give away

99% of his Facebook shares (or so it seemed) in order to set up a foundation, following the model of Bill Gates and Microsoft. My brain sprang into action, and within ten minutes I had posted a petition on change.org titled 'Dear Zuck, give us power not charity.' My ask? That instead of gifting those shares to his foundation, he might gift them to us, the users of Facebook; instead of burnishing his reputation picking and choosing which problems to spend money on, he might seize that glorious opportunity to reinvent Facebook as a truly democratic organisation.

Of course, I hadn't read the small-print in enough detail, and Zuckerberg's generosity was rather less pure than it seemed. The structure of the 'gift' would mean he retained his decision-making power, the extent of which is difficult to overstate; while he was perfectly happy giving away the money, he did not intend to give up a grain of power. In 2019, Chris Hughes, who co-founded Facebook, warned that Zuckerberg's influence went far beyond that of any government or other entrepreneur because he controlled three platforms (Facebook, Instagram and WhatsApp) used by billions of people daily:

Facebook's board works more like an advisory committee than an overseer, because Mark controls around 60 percent of voting shares. Mark alone can decide how to configure Facebook's algorithms to determine what people see in their News Feeds, what privacy settings they can use and even which messages get delivered. He sets the rules for how to distinguish violent and incendiary speech from the merely offensive, and he can choose to shut down a competitor by acquiring, blocking or copying it.[145]

Seen through a Citizen lens, this situation is both the root of the problem that Facebook has since all too clearly become, and the reason why it represents such an exciting intervention point. With this unbelievably intense concentration of power in the hands of one single, solitary man, Facebook is not even as 'democratic' as a conventional shareholder-owned company. Yet at the same time, it only needs one man to shift his worldview, to see that the idea which he birthed could grow and develop in ways that are far better for the world, and to step into the Citizen Story, for something fundamental to shift.

To be completely clear, what I am proposing is not one of the two routes to reigning in Facebook that are most widely discussed today, either breaking up the company or restricting its activities through regulation. Either might be better than where we are now, but ultimately I think they are solutions to the wrong problem: it is not necessarily Facebook that has too much power, it is Mark Zuckerberg. Starting from there, I am proposing a third approach, of shifting it into user ownership. Making Facebook, ultimately, a co-operative, and the first great Citizen Business.

Facebook has, at least in words, the perfect mission statement to unleash this energy: 'to give people the power to build community and bring the world closer together.' What might it look like to unleash the ideas, energy and resources of Facebook's 2.8 billion active users to build community and bring the world closer together? I find that a tremendously exciting question. Even at the level of the product experience, I can imagine how what is already slick could actually become much better, perhaps powered by some sort of suggestion box function, like the edit links on Wikipedia. Things could get much more interesting than that, though. Imagine Facebook were to look to BrewDog for inspiration: safe and confident in its relationship with us, its user-owners, it could open-source its algorithm code just as BrewDog does its recipes; truly committed to building community, it could create a training and development function for community organising as the equivalent of BrewDog's adoption of the cicerone qualification (perhaps replacing the huge advertising services function that currently exists). It could even move to a subscription model, or at least crowdfund new features and extensions, as BrewDog did with the Dog House Hotel.

The foundation for all this, of course, would be a governance structure in which all of us had meaningful power. As a basic first step, we would have representation on the Facebook board, perhaps through some sort of election process. But if Facebook were a co-operative, the whole company would in effect be repurposed as a genuine engine of democracy, and this would just be the very beginning. With the energy of the company genuinely focused on building community rather than selling advertising, and its 2.8 billion users applied to the task rather than just its staff, Facebook would become a hotbed of democratic innovation just as it has until now been a hotbed of advertising technology. New translation tools and video-conferencing could power a network of user assemblies, tasked with deliberating on such matters as advertising policy and the definition of news, and making recommendations to the board.

Democratic innovations that are currently taking shape slowly and hesitantly – like liquid democracy (which equips people to allocate their voice to those they trust on an issue-by-issue basis, rather than the relatively blunt nature of electoral representation) or quadratic voting (which enables the quick and simple aggregation of the preferences of large numbers of people) – would massively accelerate, with Facebook's own governance their first use case.

Facebook would not just cease to be the source of some of the greatest challenges faced by this era of global democracy; it could lead and support our governments into and through the next. And remember, lest this gets written off as idealistic nonsense... one man, who desperately wants to be loved and revered, could decide to trigger it tomorrow.

7. CITIZEN GOVERNMENT

Taiwan might seem like an unlikely place to find the world's leading Citizen Government.

The country was one of the last outposts of the Subject Story, spending most of the 20th Century under authoritarian rule. It gained freedom of the press only in 1988, and had its first presidential election in 1996. When the Consumer Story did arrive, it landed hard and fast, the capital Taipei rapidly becoming one of the world's most technologically advanced and fast-moving cities. Even as this transition took place, the old cultural norms remained lodged, expressed in adages such as 'obedience is the foundation of responsibility.' When Taiwan's right-wing government launched its 'Economic Power Up Plan' in 2012, a fusing of Subject and Consumer logics seemed very much on the cards. The communications effort conveyed the message that economics and governance are extremely complex matters, best left to the experts: the government. 'Instead of wasting time talking about policies,' the television advertising campaign said, 'let's focus on doing the groundwork to improve the economy and just get things done!' Government would be the central brain steering the market; the people would have what they wanted, as long as what they wanted stayed within the boundaries of what the government decided should be possible.[146]

On the surface, it seemed the message was accepted. But just beneath – and all around – times had changed.

2011 was, after all, the year the whole world cracked open. It began with the eruption of public anger in Tunisia, triggered by the death of the street vendor Tarek el-Tayeb Mohamed Bouazizi on 4th January, two weeks after he had set himself on fire in despair in response to the confiscation of his goods. The waves of protest and almost unthinkably immediate regime collapse spread from Tunisia to Egypt, Yemen, and Syria. In September, there were crowds in Manhattan's Zuccotti Park chanting 'We are the 99%.' On 1st October, thousands marched and hundreds were arrested on Brooklyn Bridge. More waves of protest spread out from there through cities across the West.

Occupy and the Arab Spring were just the most famous up-heavals of that astonishing year. In March, an earthquake off the coast of Japan threw a 14-metre tsunami over the seawall of the Fukushima nuclear plant; a nation with no real history of mass protest turned out in huge numbers. In July, a million Israelis took to the streets in the 'Social Protest.' In August, London burned for a week as protests following the police shooting of Mark Duggan, a 29-year-old black man, escalated into widespread riots and looting. Christmas saw upheaval in Russia added to the list, with the protests of the Snow Revolution taking hold in response to Vladimir Putin's declaration of intent to run for a fourth term as President. In the words of British journalist Paul Mason, this was the year it was 'kicking off everywhere.'[147]

In Taiwan, emboldened by what had happened around the world in 2011 but knowing that the conditions were not yet in place for mass protest at home, a group of programmers and developers began to organise. Rather than hit the streets, they set out to 'fork the government,' creating parallel ('forked') websites to those of many government departments, making whatever information they could hack and scrape together accessible and visually engaging. They showed – in deeds not words – exactly how civic participation in public decision making could operate, if only the government was open to it.

The group called themselves Gov Zero after the replacement of 'o' with 'o' in their otherwise identical forked URLs (their domains all end in '.gov.tw'). Their first project? A citizen audit system that allowed anyone to grade and comment on every budget item on a parallel site to the official portal of the Accounting and Statistics Office. Active user numbers were by no means enormous, but a growing core of students, academics and activists began to take notice. The Gov Zero community started to expand. And most importantly, there was a tangible manifestation out there of a very different idea of the relationship between the government and the people. The conditions that had been lacking were now being created.

On 17th March 2014, the moment came. The ruling Kuomintang Party (KMT), by now clearly working closely with the Chinese Communist Party, attempted a unilateral move to approve a new trade deal with mainland China without the usual clause-by-clause assessment in parliament, on the basis of continuing the 'Economic Power Up Plan.' At around 9pm the next evening, approximately one thousand protesters gathered outside the Legislative Yuan, Taiwan's parliament building. Most were students. A local florist distributed sunflowers, a symbol of hope in Taiwanese culture. What followed would become known as the Sunflower Revolution.

Drawing directly on the tactics they had watched around the world in 2011, the protesters began to climb the fences. Around three hundred made their way inside the Yuan building, occupied the floor overnight, and successfully repelled the police who attempted to evict them. Once established inside, they organised themselves into discussion groups, and began to debate the detail and potential consequences of the trade deal. Members of the Gov Zero community set up a broadband connection, and footage of the discussions spread rapidly around the country, first through social and then broadcast media.

The occupation continued for days, with the numbers of protesters growing, and universities and civil society organisations announcing their support. Critically, the Speaker of the Yuan, Wang Jin-pyng, took a position that validated the protests. Despite being a longstanding member of the KMT, Speaker Wang refused to meet with the President and Premier, arguing that they should listen to the concerns and allow scrutiny of the trade deal. He argued, indeed, that the occupation of the Yuan represented democratic space being used for democratic purposes; that it might look different to the usual operation of parliament, but it was democracy in action. On Sunday 6th April, Speaker Wang issued a promise to the protesters that no trade deal would be legally signed without full parliamentary scrutiny. Next morning, the occupiers held a press conference in which they promised to leave the Yuan at 6pm three days later, allowing parliamentary business to resume while maintaining pressure from outside. They left exactly on time, having cleaned the chamber from top to bottom.

Despite heavy diplomatic pressure from China – each time repelled with protest – the trade deal was never signed, and the government was humiliated. But that was just the beginning. In late 2014, there were local and mayoral elections around the country. In almost every instance, politicians were elected who had stood up for the protestors, even where such a result had not been remotely expected. Several had not prepared acceptance speeches when the results were announced.

At that point, central government had to take notice. A number of the country's most senior politicians reached out to the Gov Zero community, and asked for their help. In December, Jaclyn Tsai, a highly influential Minister, approached Audrey Tang, one of the Gov Zero leaders, and asked Audrey to become her mentor. The two began work almost immediately, finding ways to bring the Gov Zero approach into the actual processes of government. By the end of 2015, they had not only led the creation of a participatory policymaking platform called vTaiwan, built on the basis of an open-source software program called pol.is; they had also proven its power with a first use case of international significance, effectively crowdsourcing what today still stands as the world's most successful regulatory framework for Uber and other platform businesses.

All this happened under the same President, but the election was now approaching. 2016 brought a further change with the election of Tsai Ing-wen of the Democratic People's Party, and Audrey Tang was among the first people the new President wanted to speak to when she took office. A non-political appointee, Tang moved from the status of mentor to become Minister-without-portfolio in her own right on 1st October 2016 – and again brought the Gov Zero movement with her. 'My existence,' she said in an interview at the time, 'is not to become a minister for a certain group, nor to broadcast government propaganda. Instead, it is to become a "channel" to allow greater combinations of intelligence and strength to come together.'[148]

Tang quickly became internationally respected, as Taiwan rose to the top of the global rankings for open government and successfully ran extensive participatory budgeting processes. 'We don't care that much about whether people trust the government or not,' Tang wrote in an email to the British news outlet Tortoise, 'but we care a lot about the government trusting its people.'[149] A truly Citizen approach to government was taking shape.

CITIZENS x COVID

Then came the pandemic.

On 31st December 2019, the Wuhan Municipal Health Commission posted a public briefing on its website about a pneumonia outbreak of unknown cause spreading through the city. The news was picked up on Chinese state television and newspapers, with the *People's Daily* stating that 'the exact cause remains unclear and it would be premature to speculate.' It was also picked up by Reuters, and by outlets around the world including Deutsche Welle.

Much of what happened next, and continues to happen, we all know all too well. But very few people know anything about what happened in Taiwan; and everyone needs to.

Taiwan's government built a response on the unwavering foundational belief that its people were intelligent, capable, and responsible; individuals who had the desire and the capacity to shape the places they lived for the better. A Citizen response.

Three principles shaped its approach.

The precautionary principle was top of the list, preventing any great and plausible harm (even if unlikely) the first concern. There had to be some awareness of the economic cost of the actions to be undertaken, of course, but the government saw the nation as a society first and a market second. People were creatures of intrinsic worth first, and producers and consumers, units of economic value, a distant second. As a result, acting fast – rather than attempting to calculate exact probabilities and economic impacts – would be the default.

Next came involvement. Understanding its Citizens to be intelligent, capable and responsible, the government saw no need to protect them from unpleasant information as if they were small children, on the basis that they might panic; that would not only be patronising but also deeply wasteful of the opportunity to let everyone do their part and contribute their wisdom and experience. Instead, every opportunity would be sought to tap into their ideas, energy, and resources.

Finally, equity and fairness. Since every human being is equally worthy of basic rights, possessing of fundamental human dignity, the government sought to give everyone the same respect, the same power, and the same opportunities.

Precaution, Involvement, Equity. Audrey Tang was right at the heart of both the substance and the communication of this response, with one of her contributions being to bring some alliteration and joy to the plan, coining these principles more memorably as 'Fast, Fun, Fair.'

Acting fast, then, the government got started straight away; not over-reacting, but definitely reacting. At this point, still in December 2019, the reports were confined to the city of Wuhan, the case numbers were not high, and it could have been an existing virus. Taiwan started testing arrivals just from that city while they were still on their flights, for as many different viral strains as they could. They also notified anyone who had arrived from Wuhan in the previous 14 days, and asked them to come in for tests as well. Anyone with symptoms was asked to quarantine at home for 14 days with medical support.

Then the government started preparing the contingency in case of the worst case scenario, putting a central command centre on standby, notifying ministers and others who would be transferred to its operation. All this was in place within five days, by 5th January. The nation was ready and watching.

By 9th January, the signs were not good. Singapore, as well as several cities in mainland China, had reported likely cases. That day, Chinese state television announced the discovery of a new coronavirus, with scientists having ruled out all known viral strains. On the 11th, the first death was reported. The world, not just Taiwan, now knew this was a new disease, and that it could be fatal. The big outstanding question was whether human-to-human transmission was happening. If it was, this new disease had pandemic potential, and cases were likely to be far more widespread than had yet been accounted for in our globalised world. When confirmation of human-to-human transmission came on 20th January, it triggered the immediate decision to launch Taiwan's command centre and a major public programme.

With a senior minister in charge, the new command centre started working around the clock, coordinating action and drawing ideas from across government departments. It ran its first daily press conference that very day, to announce its launch. As Wuhan went into lockdown on 23rd January, Taiwan temporarily stopped exports of personal protective equipment from the country, until such time as it knew it had what was needed if things got really bad. It simultaneously expanded domestic production of PPE.

With all this in place, it was time to consider the other two principles: involvement and equity, aka fun and fair. Accepting the virus had almost certainly arrived, Taiwan's ambition was to make the response as much of a team effort across the nation as possible.

Believing that people want to be useful, to be able to contribute, the government saw its job as structuring the opportunities for people to do so. All of them.

It set up a national telephone hotline with a message recorded by the President herself, which any Citizen could call to propose their ideas for the national response; then the daily press conference was used to celebrate these contributions, making them famous and encouraging more. Among the callers was a six-year-old boy, who called up concerned that his classmates did not want to wear their regulation-issue pink face masks, asking for the nation's baseball team to make them more appealing. Six of Taiwanese baseball's biggest names joined a press conference, resplendent in their pink masks, just a few days later.

Alongside this mass broadcast effort, the government started setting open source software challenges for developers, building on the established Gov Zero capability, and knowing that apps to help share information about the spread of infections, as well as to manage distribution of resources such as face masks and medication, could be a real boon. Tens of different applications and tools were developed, some at the government's request, others simply through the energy of the community.

Efforts to combat misinformation were similarly handled. The government published each bit of social media misinformation they found, challenged the population to 'out-meme' it, and supported the correct information to spread.

Then, as it became clear by early March that the government's fast action meant Taiwan had escaped the worst of that first wave, they turned out to the world, offering their help where they could, publishing – as a first contribution – a list of 124 actions they had taken, in an international journal, in English. By April, with that first wave successfully contained, the government turned to designing incentives to give people confidence to get back to supporting local businesses and back to work, in the form of direct cash payments to all citizens.

As of the end of April 2021, Taiwan had recorded only nine coronavirus deaths and fewer than 1,000 cases, despite recording its first confirmed case on 21st January 2020, without ever going into lockdown. Not only did the Taiwanese weather that first shock, but they had since been able to spot and control two subsequent outbreaks almost immediately. And with the health of citizens secure, the Taiwanese economy entered 2021 in fine fettle, forecast to grow at its fastest rate in seven years.

This is what is possible with Citizen Government.

GOVERNMENT BY PROTOTYPE

Taking the Covid response out of context, it may be tempting to dismiss Taiwan as an anomaly, different in a way that means we couldn't ever have expected any other nation to follow its lead, too small to deserve focus. It's perhaps true that Taiwan was uniquely likely to respond with suspicion and speed to a crisis emanating from mainland China: the Chinese Communist Party has never accepted Taiwan's existence as an independent state, and has used its global power to enforce Taiwan's exclusion from bodies such as the United Nations and the World Health Organisation; Taiwan was also hit hard by SARS in the early 2000s. And it is indeed a relatively small country, with a population of 23 million, three million in Taipei, the capital. It's tempting to think like this, it really is, because Taiwan's story confronts us with the possibility that millions of people worldwide and well over a hundred thousand here in Britain didn't have to die. That's a deeply painful truth to face.

I believe we need to have the courage to face it, however. We need to accept that China's geopolitical threat has long been clear to all, and that all countries knew what Taiwan knew, when they knew it, through the early months of 2020. We need to acknowledge that New Zealand, a country culturally very similar to the UK, explicitly adopted 'the Taiwan model' as early as 15th March. We need to remember that Taiwan is not so small after all and indeed has more than twice as many people as Sweden, a nation far more discussed during the first wave of the pandemic, and almost as many as Australia. This isn't about blaming any one individual or political party. It's about learning, and changing. If we can face these truths, our reward will not only be a view of how things could be different, both in daily life now and indeed when (not if) the next crisis hits. It will be far greater than that. We will be able to see a way forward for a political system that is arguably in meltdown, and the promise of rapid renewal. Taiwan's turnaround from the brink of a return to Subject Government in 2012 to Covid-ready Citizen Government by 2020 shows us what government looks like in the Citizen Story; it also shows us how we might get there, and promises us the journey could be fast.

Tracking the ideas of the last two chapters, part of the lesson is about the need to rearticulate the purpose of government. In the Subject Story, governments represent the God-given elite who can and should tell us what to do because they know best. In the Consumer Story, government becomes just another service provider. In the Citizen Story, the purpose of government is to provide the space and the means for us to come together to meet our collective needs, be they urgent and immediate or ongoing and sustained – but stopping well short of doing it for us. The triangle from Chapter 5 applies: the purpose of government is to enable people to make our society better ourselves, not to do it for us, or to us.

This in turn demands that the institutions of Citizen Government act as platforms not deliverers, sharing power and inviting participation just like the businesses we met in Chapter 6. They must be neither dominating nor subservient, but active alongside citizens, putting forward suggestions and resources and opportunities for us to act; they must be neither in control nor in service but in support, equipping us to build our own future together. I could populate a whole set of the Seven Modes of Everyday Participation with examples of emerging government practice from around the world, all of which stand in stark contrast with current norms of election, consultation, and little else. There's Calgary Mayor Naheed Nenshi's '3 Things for Calgary'[150] campaign for the Share Connections mode, for example, which saw Nenshi and his team flip the usual tired rhetoric of local politicians 'working hard for you' and instead challenge every Calgarian to state three things they would do for the sake of the city, and nominate three friends to do the same. For the Contribute Ideas mode, there is Better Reykjavik, the initiative that allows citizens of Reykjavik to put forward their own proposals for how the city could be better, vote on each other's, and see those ideas debated in a special monthly session of the City Council. Or 'civic crowdfunding' platform Spacehive, which works in partnership with local authorities all over the UK to match fund projects put forward by local people....[151]

The most important lesson of Taiwan, though, is not so much about the concept of government expressed – its purpose, or its role as a platform – but about the process by which this rapid transformation has been achieved. This has been a process of prototyping, of iteration. Instead of getting stuck in the theory or the search for some kind of utopian switch we can flip to change everything all at once, Taiwan illustrates starting, sharing the challenges, inviting people to get involved, taking action to-gether, understanding impact together, building the energy, and going again, and again, and again. And it has been a process that is itself focused on the processes rather than the institutions of government, on what government does rather than what govern-ment is. Indeed, what Audrey Tang, Wang Jin-Pyng, Tsai Ing-wen, the Gov Zero community and the entire Taiwanese nation have been building is less about government, and more about gov-ernance – less about the institutions of government at all, and more about the processes by which we make the decisions of our society together.

In Taiwan, this approach began outside government back in 2012, when the Gov Zero community used what they had to build what they could, picking the Accounting and Statistics Office first, and demonstrating what a Citizen version would look like in practice. It continued in the protests of the Sunflower Revo-lution, as the students inside the Yuan building practised a new model of deliberation during the occupation, and broadcast it to the nation. And when the Gov Zero movement was brought into government in response to the protests, the prototyping spirit came with it, with the best example being the development of the regulatory framework on Uber in 2015.

It is worth reflecting briefly on what a conventional, linear policy development process would have looked like – indeed, on what conventional processes all over the world on this issue have looked like. Seeing it as the role of government to deliver a regulatory framework, a group of civil servants would have been tasked with finding the answer. These would have become the target for extensive lobbying from interest groups during the drafting process (the best funded and most persistent often being the biggest corporations), and would have worked closely with the elected politicians who would need to champion the output. This would create a great deal of tension and argument, slowing the process down. Eventually, a 'draft' proposal would then have been published for 'consultation' – but so much effort, time and money would have been invested by that point that it would be more a *fait accompli* than a draft, the so-called consultation really more an exercise in selling an output to people-as-Consumers than inviting input from people-as-Citizens. The public would know that, and would not respond. Their silence would be taken as apathy, and the proposal would likely go through. It is possible this might have got to the right answer – but the evidence from the continuing struggles of urban administrations around the world to deal with Uber, as with so many other challenges, suggests otherwise.

What happened in Taiwan was completely different. Instead of disappearing to find the answer, the starting point was to share the question: 'How can we regulate Uber so that we maximise the

benefits and minimise the negatives?' Instead of pretending interest groups didn't exist, these were explicitly identified – Uber
drivers and passengers, conventional taxi drivers, and so on –
and invited to be part of the process. Then the vTaiwan digital
platform came in. Using this tool starts by asking participants
to write simple, short statements expressing their view of the
situation – from 'Uber should be banned' to 'There's no need to
do anything' – and then asking them to agree or disagree with the
statements put forward by other participants. The tool generates
a map clustering participants into groups by their responses,
letting them see how divided they are. Then comes the clever bit:
from this point, participants essentially compete to articulate
statements that can cross the divides, iterating until something
approaching consensus is found. In this case, the threshold set
was 80% agreement across the whole group, and eventually
seven statements emerged – things like 'Uber should be taxed
as a transport company' and 'Uber drivers should be considered
employees.' I say eventually; the whole process took four weeks
from start to finish, and all government officials had to do was
host the conversation.[152] Share the challenge. Start prototyping.

By 2018, the prototyping approach had become structured
into the institutions of government, with its clearest form being
the first Presidential Hackathon. Focused domestically in its first
year, but since expanded to welcome teams from across the world,
this process begins with the President setting out an annual
challenge rooted in the United Nations' Sustainable Development
Goals ('Enabling Sustainable Infrastructure' in 2019, 'Using data
to enable the SDGs' in 2020). Applications then open, before the

teams putting forward the most promising ideas are invited to Taipei for a three-day intensive working session to develop their proposals with the support of Taiwanese civil servants. The fourth day is Demo Day, with funding announced for the projects that will be taken forward to implementation. It is a process 'borrowed' directly from Silicon Valley prototyping approaches to building billion-dollar businesses, harnessed in service of domestic and global civic challenges.

When Audrey Tang coined the term 'Fast, Fun, Fair' for Taiwan's Covid response, then, she was really describing the nation's whole approach to government. Swift protective action came first, keeping people safe the first and most basic condition for participation. Then came the framing of the response as a national team effort, made real in the telephone hotline, the open source software challenges, and the collaborative, iterative approach to combating misinformation. These tools and processes served to focus Taiwanese citizens' efforts, equipping them to make their collective impact add up to more than the sum of their parts. As the task turned to economic recovery, a small-scale test stimulus was launched, using small cash payments direct to citizens rather than doling out relief to selected supplicant sectors and organisations. When it worked, its scale was increased. Share the challenge, act together, learn what works, then go again....

INCH WIDE, MILE DEEP

It is powerful if somewhat depressing to think about Bianca in light of what is happening in Taiwan; she could all too easily be Germany's Audrey Tang if only there were a Speaker Wang to open the door. But while Taiwan is to date the only nation in the world that has truly embraced Citizen Government, a big part of prototyping is starting small, at a scale where doing things differently is possible. That's exactly what's happening all over the world; it's what Immy and Kennedy are both part of; and there is an excellent example taking place in the east London Borough of Barking and Dagenham, just a few miles from where I live.

As with Taiwan, Barking and Dagenham is not an immediately likely location for Citizen Government. The population of the borough, currently a little over 200,000, is both fast-growing and rapidly diversifying. Roughly 80% of residents were white British in 2001, with that figure dropping to below 50% by 2011. For four years, from 2006 to 2010, the openly racist British National Party was the official opposition on the Borough Council, holding 12 of 51 seats. Yet it was here that in March 2016, shortly before Britain voted to leave the European Union, conversations started between Chris Naylor, Chief Executive of the Council, and Tessy Britton. Tessy is not a fan of labels, but when pressed describes herself as a 'social designer.' By this point, she had already been running 'Participatory City' experiments in various London boroughs – with interventions like co-working spaces, tool libraries, food growing and tree planting – for nearly a decade.

Over the course of the next year, Tessy not only managed to inspire Chris (and help him inspire his elected Councillors) to host something much more ambitious, together they raised enough funding to get things started. In August 2017, the Every One Every Day project launched with a stated ambition to offer everyone in the borough opportunities to participate, in the process sparking at least 250 new community projects and 100 community businesses, within five years.[153]

Tessy explicitly saw what she was trying to create in the language of a platform – she calls it a 'support platform.' In her model, there are two interlocking systems, depicted as a kind of vertical Venn diagram. The upper system, above the surface so to speak, represents the proliferation of projects, initiatives and businesses, started and run not by Tessy's team, but by citizens themselves. The second system is the support platform, and is made up of three main elements. First, a pooled operations and logistics function set up to offer administrative support to all those projects. Then a set of five physical spaces, four high street 'shops' which offer highly visible access points and a larger warehouse which acts as a sort of open, shared office-cum-factory. Tessy and the team describe this as a Public Makerspace, a key piece of infra-structure that gives open access to spaces, tools and equipment for making, repairing, learning, and sharing. The third, and perhaps most crucial element of the support platform is a small but highly skilled team of open-minded, yes-we-can facilitators, trained to work with, draw out and enable the ideas of the community.[154] (Tessy stole a brilliant member of the New Citizenship team, Iris Schönherr, to be part of this group. I will get over it.)

This structure in turn unleashes a wave of prototyping, iter-ating, and building. The facilities in the warehouse manifest this physically: there are 3D printers, kitchens, and gardens made available to local people to develop prototype products that might become outputs of new community businesses. The team refer to these simply as 'first versions,' demystifying and translating one of the key concepts of a prototyping approach – Silicon Valley's 'minimum viable products' – into normal speech.

It's working, with thousands of the citizens of Barking and Dagenham now actively involved in already well over 200 projects, ranging from breakfast clubs to upcycling workshops to street ball games, all organised and led by local people. A good number of them are well on the way to incorporating as profit-making community businesses. The big dream is to take the pressure off existing Council 'services,' from adult and child social care right through to something as prosaic as waste disposal. This last in particular may seem surprising, but only needs a moment's thought to understand: when people feel they have meaningful agency in the place where they live, they look after it better. Perhaps most excitingly, the approach is beginning to spread (rather than scale, as Silicon Valley might aspire to): the warehouse has welcomed visitors from across the country and indeed the world, many of whom are taking the ideas and principles back home with them and getting started in their own neighbourhoods, towns and cities. Remember the idea of 'inch deep mile wide change that schisms the existing paradigm,' articulated by adrienne maree brown and beloved of Immy? What Tessy and the team are doing in Barking and Dagenham is a pretty good illustration of what that looks like in practice.

MAKING THE BIG DECISIONS FASTER, TOGETHER

Tessy Britton and the growing movement of participatory 'shops' and warehouses speak to a big part of what Citizen Government is all about, the day-to-day experience of organising and meeting needs and doing life together. But national (and supranational) governments in particular also have a legislative role: they have to set the laws which act as the parameters for all this activity. This might seem less obvious territory for a Citizen approach and for the spirit of prototyping – but actually, there is a wave of democratic innovation building in exactly this space that is hugely exciting, and is gaining traction even in the institutions that were originally created to carry the Consumer Story. The Organisation for Economic Co-operation and Development (OECD), for example, founded in 1961 to stimulate economic progress and world trade, now has a major workstream dedicated to open governance, and fosters a practitioner community called the Innovative Citizen Participation Network, of which I'm a proud founding member. The central focus of the network is deliberative democracy, and in particular Citizens' Assemblies.

The decision of the Irish people to legalise abortion, confirmed by a two-thirds majority at a national referendum in 2018, is arguably the best example of this process in action. With a total ban on abortion having been part of the constitution since 1983, the Assembly was originally commissioned by a government that acknowledged itself to be deeply stuck on the issue, with parties on either side of the political spectrum concerned about alienating their 'bases' with any form of compromise. The solution was to return the question directly to the people, not by going directly to referendum, but by creating a 'mini public' at a scale where it would be possible for them to work together intensively to learn about every aspect of the issue, deliberate on the options, and make a legislative recommendation back to government.

Ninety-nine Irish citizens were accordingly selected, initially invited at random but then filtered to create a representative sample of the national population.[155] Over five weekends spread out over five months, the Citizens' Assembly shared stories in small groups, received evidence from a balanced panel of 40 expert witnesses, heard directly from women affected by the laws, and from 17 campaigning and lobbying groups. All proceedings were fully transparent to the public and media, and the Assembly received a great deal of coverage. As per politicians' expectations, there was suspicion from both sides: campaigners who believed the ban on abortion should be upheld argued that the process was fixed; when it emerged that early Assembly deliberations were leaning towards revising the conditions of the ban rather than repealing it outright, opposing campaigners also railed against it.

Ultimately, however, the Assembly recommended a referendum to repeal the eighth amendment, with 64% of Assembly members voting to introduce 'terminations without restrictions.'[156] Still there was concern: this proposal was initially seen as too extreme and out of step with 'Middle Ireland,' and there were significant and understandable fears about launching a referendum process in Ireland in 2018, a country and a time deeply affected by the ongoing reverberations of Britain's decision to leave the European Union.

The Assembly's recommendation was adopted, however. When the campaign began, it turned out the preceding steps had made the referendum 'immune' (in the words of the political commentator Fintan O'Toole[157]) in the face of the disinformation, interference and manipulation that had affected not just the Brexit process but so many elections and referenda in recent years. Why? Because actual people across the country could say: I was involved in the process, and that never happened, they're lying. And, of course, the people of Ireland did agree: 66.4% voted for repealing the eighth amendment, removing the ban on abortion, and legalising unrestricted access up to 12 weeks.

In the event, the sharing of both expert evidence and people's stories forced the country to face up to the fact that abortion was already prevalent in Irish society. It also generated greater empathy on both sides for the feelings, experiences and perspectives of others. Many 'Yes' campaigners leaned into the new space of empathy opened up by the Assembly process by engaging everyone they could, rather than writing off typically conservative sectors of society as a lost cause. As O'Toole explains, 'It turned out that a lot of people were sick of being typecast as conservatives. It turned out that a lot of people like to be treated as complex, intelligent and compassionate individuals.'[158] A majority of farmers and more than 40% of the over-65s voted Yes.

It might initially seem odd to describe Citizens' Assemblies and other similar deliberative processes as part of a prototyping approach to government. We tend to think of prototyping as about a small group of people getting into action, not the talking and taking of time to digest that is the essence of deliberation. But I think the Irish example makes the point powerfully that this is exactly what deliberative processes are about, because it reminds us of the contrast with the alternative, conventional approach. Five weekends, a 'mini public' of unusual suspects, a solution to a challenge on which government had been stuck for nearly 30 years? Sounds a lot like a prototype to me. From this perspective, Citizens' Assemblies are basically civic hackathons.

YES WE CAN

Towards the end of April 2021, Taiwan's Covid defences were finally breached. Two China Airlines cargo plane pilots, infected with the virus, were theoretically in quarantine at Taipei's airport hotel, but in the context of increasingly relaxed restrictions, they somehow came into contact with other guests. Very soon, cases were multiplying, and by the end of May, the total number since the start of the pandemic had more than tripled. By this time, roughly half the UK adult population had received their first vaccine dose, and a third had received both; China had vaccinated over a third of its population, dispensing half a billion jabs. By contrast, only around 1% of Taiwan's much smaller population had been vaccinated even once.

International media outlets raced to tell the story, paying Taiwan significantly more attention now (*schadenfreude*?) than at the onset of the pandemic. 'How a false sense of security, and a little secret tea, broke down Taiwan's Covid-19 defenses' ran the *Time Magazine* headline, focusing attention on the role Taiwan's 'tea houses' (essentially legalised brothels) had played in spreading the outbreak.[159] 'A victim of its own success: how Taiwan failed to plan for a major Covid outbreak' was the *Guardian*'s counterpart,[160] just as crowing if a little less salacious. The lesson, they posited, was that Taiwan had been on borrowed time throughout, beating Covid more by luck than judgement, and only temporarily at that. More than that, the limits of its

Citizen approach were now coming back to bite it. With vaccines the only real solution, relying on public behaviour rather than mass procurement (the UK's strategy, per the Consumer playbook) or rapid rollout (only possible in China's highly ordered and efficient Subject society) might have been nice, but ultimately failed in the harsh light of a sustained crisis. Taiwan was unvaccinated and exposed.

What are we to make of this? Is the Citizen Story nice, but ultimately not up to the job? Ultimately, is the viable contest for the story of society only really between Subject and Consumer?

No. The numbers above may be accurate. But China's role is far darker, Taiwan's travails far less serious – though still deeply tragic – and the picture in the UK far more nuanced than this surface version of events would have us believe.

To take these in turn, there is, first of all, a very good reason why Taiwan's vaccination programme was so limited. Before the May outbreak, Taiwan's only significant access to vaccine supplies was via China, which not only controls its own Sinovac product directly, but also has indirect control of the availability of other vaccines in the region, via the Chinese companies which act as agents. The Taiwanese government understandably feared that the Chinese Communist Party might seek to use vaccine contracts as a way to access both government data and the private data of its citizens. It is a fear which seemed to be justified in mid July, when BioNTech's Chinese sales agent put forward a contract which would have given the firm access to Taiwanese medical records. Only after this did the United States and Japan involve themselves in the provision of vaccines directly to Taiwan; the

German government has also found itself a role in BioNTech's contract negotiations.

Secondly, despite the implications of the headlines, and despite its significant scale relative to anything the country had yet endured, the Covid outbreak in Taiwan in May 2021 was by global standards both vanishingly small and rapidly controlled. By the end of July, new cases were back down below a dozen a day from a single-day peak of 670, with a sum total from the whole pandemic of a little under 16,000; 777 people had lost their lives in addition to the nine who had died up until the end of April 2021.[161] There had still not been a full lockdown on the scale of those seen elsewhere in the world, even in Taipei, the epicentre of the outbreak.

Contrast those numbers with the UK. On 24th January 2021, 1,820 deaths were recorded on a single day, significantly more than double the Taiwanese total from the entire pandemic. By the time this book went to press in winter 2021, Britain was climbing towards ten million total cases, and staring down the barrel of another tragic winter. The assertions that Taiwan's defences were fundamentally broken, or that it had 'failed to prepare,' are questionable at best.

Finally, the perception that the UK's vaccine success is a manifestation of the Consumer Story as ultimately the way to get things done – down to 'greed and capitalism,' as Boris Johnson put it – is at least as questionable. This is the case both in terms of the procurement and the rollout. The Vaccine Taskforce, which led the procurement exercise, sat outside the normal operations of government, and ploughed huge amounts of public money into vaccine development in the form both of investment and purchase contracts, working closely and directly with universities and pharmaceutical companies. This was a smart decision, but quite the opposite of classic Consumer logic: this was the UK government stepping into its agency to drive the market in the interests of its citizens, not stepping back to allow the invisible hand to function unfettered. The motive was not profit, but the purpose of serving public health. It is better understood as an isolated departure from the dominant logic of the UK government in the pandemic than its clearest expression. As to the rollout, to the extent that this has been a success, credit is due primarily to Citizen action, to the efforts not only of National Health Service staff working in distributed local organisations, free of their normal constraints, but also to over 100,000 volunteers whose contributions have been critical to enabling its speed; many of these first signed up as NHS First Responders in April 2020. Where the rollout has begun to falter, the Consumer Story is arguably to blame: the removal of restrictions in mid July 2021 in a bid to boost the economy saw a sharp drop off in vaccination rates, with many receiving the message that the danger was over.

As things stand at the end of November 2021, Taiwan's spring outbreak is a distant memory. There has been no further spike, and a higher proportion of Taiwanese than British have had their first vaccination, with a new domestically produced vaccine listed among the options available to its citizens. Its economy is thriving, and applications for the fourth annual Presidential Hackathon are open. Citizen Government doesn't sound so bad to me.

CITIZENS' BRITAIN

Britain may not at first glance seem the likeliest place for the Citizen Story to gain a foothold in the West. After all, the Johnson government's response to the pandemic has veered from Consumer to Subject and back again – think 'herd immunity,' 'Eat Out To Help Out,' and so on – with barely a sight of Citizen thinking in the mix. Similarly, the Brexit-driven project of 'Global Britain' takes its inspiration from Consumer dreams of unfettered markets and Subject-derived nostalgia for empire. Indeed, in *Britannia Unchained* – a 2012 book by five Conservative MPs, of whom four are now ministers – the authors make their lack of belief in their people completely explicit, stating that the British are 'among the worst idlers in the world,' that 'too many people in Britain prefer a lie-in to hard work.'[162] The belief in the capability and creativity of the British people that would be the essential foundation of Citizen Government seems to be sorely lacking.

Yet I have made the same observation about both Taiwan in 2012 and Barking in 2016. The starting context appeared auspicious in neither. And despite surface appearances, the underlying conditions are beginning to align.

First, for all that some politicians might not share it, the belief of the British people in one another has grown significantly during the pandemic: according to data from the global thinktank More In Common, the percentage of the British population saying they feel part of a community who 'understand, care for and help each other' rose from 49% before Covid to 57% in summer 2020 to 63% in early 2021.[163] A similar shift has occurred in the numbers who feel empowered to make things better in their own communities;

and over the same period the sense of Britain as a divided nation has declined significantly. The British people, in other words, are ready and willing to engage as Citizens. Whether under existing banners like Extinction Rebellion and Black Lives Matter or in new and different forms, this will inevitably manifest in a wide range of campaigning and activism as the pandemic recedes and energy levels are restored.

Meanwhile, this build-up of Citizen intent is beginning to be recognised and understood in formal politics, and across the traditional left-right spectrum. Inspired by Biden's victory over Trump, the Labour Party and the Liberal Democrats are taking the time in opposition to rediscover the roots they have in the community, investing significant effort and resource in building local power (pushed along by the Greens and the emerging 'Flatpack Democracy' movement of independents)[164]. At the same time, a recent report by Onward, an influential Conservative thinktank, set out an agenda for *The Policies of Belonging* rooted in putting power and resource in the hands of individuals and communities. 'Our aim,' the report's authors write, 'is to give citizens and their communities the power and resources to shape their places and meet the needs of their members. At [this report's] heart is a simple idea: that people are pro-social given the right conditions[165]....' It's a perfect articulation of a Citizen agenda. Another thinktank, Demos, is working with the Cabinet Office to test out the pol.is software program that underpins the vTaiwan policymaking platform. County and District Councils led by all political colours are working to unleash the power in their communities, especially having seen that potential express itself so clearly in the mutual aid groups and vaccine rollout networks of the pandemic. The New Citizenship Project is working with several of them.

What would it look like, then, to build up this prototyping, Citizen approach to the point where it is ready to become the governing story in Britain, as it has in Taiwan? I think there are three initiatives emerging, perhaps even movements, that represent the key pillars. If those involved in each can embrace this prototyping spirit, and keep growing the energy around their work, it will only be a matter of time.

The first, and most foundational, is the movement for a Basic or Guaranteed Income.

The reason this matters so much is that introduction of this measure would fundamentally shift the understanding of absolute poverty in the UK, from a personal choice for which the individual is responsible and must change himself, to a policy choice, a collective responsibility which we must change together. By contrast with existing welfare provision, it would express a deep belief in people. Existing approaches are built around conditions that must be fulfilled in return for 'benefits': the underlying assumption is that the recipient cannot be trusted, must be checked on to make sure they are not perpetuating the bad choices through which they created their own situation. Basic Income approaches are the opposite: the underlying assumption is that everyone can and wants to make a contribution to society, and that poverty erects a fundamental barrier to that contribution that must be removed not just for the good of the individual, but for the good that individual will then contribute to society.

The movement towards some form of Basic Income is growing rapidly in the UK through what is very much a prototying approach. The smaller political parties are starting to adopt it as policy, a geographic network of UBI (Universal Basic Income) 'Labs' has sprung up across the country, and there is good contact between the UK movement and its US counterpart, Mayors For A Guaranteed Income, which has now gathered pledges from over 50 US Mayors to at least trial the approach.[166] The idea has majority backing from the UK public whenever polling is undertaken, and the semi-independent governments of both Scotland and Wales are committed to trials of their own to begin imminently.[167] This is what prototyping can do: soon, this momentum will become irresistible.

The second initiative-cum-movement is deliberative democracy. Momentum behind Citizens' Assemblies is growing both locally and nationally in the UK, with the climate emergency proving particularly fertile ground for a new approach to political decision-making. Participation charity Involve has records of Assemblies held by 17 different councils on climate alone since 2019[168]; and then there is the national UK Climate Assembly.[169] This process, although not explicitly commissioned by the government and significantly smaller in scale and visibility than the French Citizens' Climate Convention, took its mandate from not one but six Parliamentary Select Committees and achieved some public note, particularly when Sir David Attenborough addressed the opening gathering.[170] And while climate is where this new momentum began, largely in response to Extinction Rebellion's efforts in 2018, it is far from the only issue that lends itself well to this approach, as is shown by the Irish abortion referendum.

Indeed, if the Irish process offers the single best example of deliberative democracy in action, it's a proposal that has been developed in Canada that I think speaks most powerfully to the potential of deliberative democracy as a carrier of the Citizen Story. Toronto-based consultancy business Mass LBP, which 'works to see that more people have a hand in shaping the policies that shape their lives,' put forward a proposal in 2019 for the creation of a 'Democratic Action Fund.'[171] The essence of the idea is that the Canadian government should set aside funding to run enough Citizens' Assemblies each year that within three to five years, every citizen would either have participated directly in one of these processes, or would know someone who had. Imagine what this would do for our understanding of what democracy is, and of our role in it. In our current Consumer Democracy, our participation is characterised primarily and almost exclusively by voting, the act of choosing between options that someone has defined for us. Mass LBP might just have come up with the perfect intervention to hack that conception, right at its core. And if in Canada, why not here?

Finally, in order for the work of UK government truly to become faster and more responsive, political power must be pushed out not just from London, but from Whitehall, the single tiny area of London where almost all the decisions that truly matter are taken. This is what is needed to unlock the potential of Tessy's Participatory City initiatives, Immy's pioneering efforts in Birmingham, and their counterparts across the country. More power, and more meaningful power, needs to be situated far closer to where Citizens are.

For all the talk of localism and devolution in recent decades, Britain remains over-centralised to an extent almost unparalleled around the world. 95 pence in every £1 of tax paid in Britain is taken by Whitehall, compared to 69p by central government in Germany,[172] for example. London is often seen as the place where the centralised decisions are made by those in other cities and towns around the country, but even London should be considered a victim of this rather than a perpetrator. Its administration receives almost three-quarters of its budget from central government, rather than through its own tax-raising powers, compared to less than a third for New York, less than a fifth for Paris, and less than a tenth for Tokyo.[173]

It is the rest of the country, though, that suffers most. Far from taxes being taken from London and redistributed to less affluent places across the country, the truth is that a huge proportion is simply being swallowed up in ineffectual central administration. The city of Sheffield, for example, receives approximately 25% below the national average in per person public spending, and even then, local government in Sheffield only controls less than 15% of how that money is spent. Simon Duffy, Director of the Centre for Welfare Reform, describes the situation as 'colonialism 3.0:'

Colonialism 1.0 was about starting new states, perhaps where nobody was living, and Colonialism 2.0 was about establishing bases in foreign countries in order to exploit those countries, often combined with military power. But I think Colonialism 3.0 begins when this system of exploitation bends back into the mother country and begins treating its own people as if they were natives of a foreign country.[174]

Duffy is also deeply critical of the nature of such decentralisation as has taken place in Britain, describing it as more performative and tokenistic than meaningful, a fig leaf to cover what is really going on: 'Mayors are better than parliaments; city regions are preferable to counties; strong leaders are preferable to councillors making decisions.'[175] All these preferred structures are in effect 'local fiefdoms,' more easily controlled by central government than the alternatives, more analogous to colonial governors than independent government.

What would it look like for this to change? It's really very simple. Power, not just money, would be distributed to the local and regional level. Take Finland, for example, where local government is funded primarily by local income, property and corporate taxes, at rates set locally. These local taxes make up 23 pence in every £1 of the total tax take in the country, compared to the figure above of 5p in Britain.[176] Central government spending then tops up where necessary, providing grants according to population and need, but is not the primary actor. This situation structures incentives and rewards for thriving local economies. With local government also having much greater power over how money is spent on things like education and social care (and far fewer unnecessary central inspections), there is also much more incentive for local people to care about local politics. The turnout in Finnish municipal elections in 2017 was nearly 60%; in England the same year, the figure was 35%.[177] A 2019 article on the BBC website was titled *Council Elections: Why don't people vote?* The answer is simply this: because councils don't matter enough.

The good news is that momentum is building on this front

too. From community wealth building initiatives in places such as Preston in the north west of England and Fife in Scotland, to the quiet emergence of Regional Mutual Banks in the south west, from growing energy among the field of 'community business' to growing calls for a Community Power Act, this movement is as yet more distributed and less clearly focused than the work towards a Basic Income or the institutionalisation of deliberative democracy – but it's certainly happening.

Ultimately, despite the obvious counter trends, I not only think all this is possible; I think it's likely. There is now an existential crisis in Britain – the tenuously United, possibly-soon-to-be-divided Kingdom. Brexit has created massive systemic uncertainty that had only just begun when the pandemic hit. Independence movements in Scotland, now Wales, and soon likely Northern Ireland increase it. The United Kingdom needs a new unifying idea and approach if it is to hold itself together.

CLOSING

There is a Citizen future emerging. As this book goes to press, a new platform called Restor has launched to allow grassroots nature conservation projects from all over the world to plot their impact, connect and collaborate. The city of Paris has just approved the creation of a standing Citizens' Assembly as part of its governance structure, and committed to distributing in excess of 100 million euros a year through participatory budgeting processes. Chile is in the midst of a Citizen-driven Constitutional Convention, which could see the country go beyond even Taiwan in its embrace of participatory democracy.

But there is another future. In fact, there are two.

'OK guys hear me out,' tweeted Delian Aspourhov on 4th December 2020. The 20-something Bulgarian is a Principal at Founders Fund, a highly influential Silicon Valley venture capital firm, and also President of Varda Space Industries, builder of the world's first space factories: plenty of people listen when he speaks. He continued, 'what if we move Silicon Valley to Miami.'[178] The following day, Miami Mayor Francis Suarez responded: 'How can I help?'

Fast forward five short months to the end of April 2021, and the inaugural Miami Tech Week was under way. Hundreds of venture capitalists and startup founders flocked to the city from all over the United States, but particularly from San Francisco; the traffic was enough to push return airfares from the Bay Area to double their usual level. Mayor Suarez was there to greet them all, the words of his tweet now the tagline of the event, rendered in vivid pink and blue against the black background of T-shirts, baseball caps and other merchandise. Over the following week, a flood of open meetups, closed-door discussions and parties were arranged and shared over messaging apps.[179]

'We're calling it South By South East,' one startup founder told *Wired* magazine.[180] The reference is to South By South West (SXSW), the influential annual gathering held in Austin, Texas, which had cancelled its 2021 events due to Covid.

If those behind it prevail, the significance of Miami Tech Week will not just be about SXSE taking over from SXSW, or even Miami taking over from Silicon Valley. The way a few people (almost all men) with a lot of money see it, it's a test run for a whole new future. It's a future that would see the Consumer Story rise like the phoenix from the ashes of its destruction, much as the Subject Story reasserted itself across Europe after World War I.

In the worldview of these people, the cracks of our time are not cracks in the Consumer Story, but in the 20th Century democratic nation-state: it is this that was a creature of its era, has always been fundamentally flawed, and must now be torn down to make way for the new. They see the state as a racket, leveraging outrageous charges on them for the right to operate; tax as protection money. They have had to put up with this until now, but things are changing, and they won't have to for much longer. Coupled with the freedom of movement that 'location-independent work'[181] enables, the next wave of technology, cryptocurrency in particular, will make it impossible for governments to track and therefore tax private transactions, destroying the business model of the state and rendering it obsolete. Nationality will soon be seen as a kind of forced allegiance; to be replaced by voluntary allegiance to communities that organise themselves 'bits first, atoms second.' First they will form online, determine rules and obligations, sustaining these efforts with their own currencies. Then they will either purchase land to make it real – as Paypal, Tesla and SpaceX founder Elon Musk is now attempting with his 'startup city' of Starbase, Texas – or they will go somewhere they are made welcome – like Miami, for instance.

This vision is as deeply rooted in the Consumer Story as was Margaret Thatcher's.

These founders consider themselves to be the cleverest of self-made men to have climbed society's ladder. They amass lists of welcoming places and ways to 'achieve financial escape velocity,'[182] including tax havens, startup visas and 'golden passports.' The benefits of technology, whether artificial intelligence, bio-, neuro- or agritechnology, will accrue to them – as will all the power in our society. In their view, their single-minded pursuit of their own self-interest should open up the best possible future for society as a whole, although the lazy and stupid majority will likely get in their own way, causing their own demise.

Many of these new 'cognitive elites, who will increasingly operate outside political boundaries'[183] are calling themselves 'Sovereign Individuals,' after the title of a 1997 book which has become a cult classic in this community, cited regularly by all the most influential players from Musk to Peter Thiel (Musk's co-founder at Paypal and a Partner at Founders Fund, where Delian Aspourhov works) to Balaji Srinavasaran (sometimes referred to in the community as 'Einstein if he had the internet.') In its pages, authors William Rees-Mogg[184] and James Dale Davidson write: 'The new Sovereign Individual will operate like the gods of myth in the same physical environment as the ordinary, subject citizen, but in a separate realm politically.'[185]

Like Thatcher, these people believe themselves to be on a moral mission. The top reply to Asphourov's Miami tweet reads 'I will be there!!!! You're doing the lord's work.'[186] This messianic belief, as with all messianic beliefs, will form the justification for inflicting utter destruction on the rest of us. It's no wonder Zuckerberg sits in congressional hearings looking vaguely nonplussed; he genuinely believes himself to represent a higher form of social evolution. And so in the future the Sovereign Individuals are working for, the Consumer Story would fuse with the Subject Story; these 'self-made' men would replace the God-given few, and become gods themselves.

THE CHINESE DREAM

Fanning the self-righteous flames of Sovereign Individuals is what they see as the only alternative: the vision of the future held by the Chinese Communist Party, and President Xi Jinping in particular. Musk, Thiel and co can see the technologies arriving which can finally set them free from the constraints of the nation state; but they can also see the Chinese government threatening to take charge of all those technologies, harnessing them to create the least free society in human history. And they're right. This is the other possible future; it's being built from within the Subject Story; and it's very much under way.

At the end of the 20th Century and beginning of the 21st, the Consumer Story was expanding into China, its final frontier, and landing in a big way. By 2005, the Chinese were outpacing Brits and even Americans in taking in advertisements and shopping.[187] For a while, with President Hu Jintao and the 'People's Premier' Wen Jiabao, it appeared the Consumer Story might reform Chinese politics as well.

But then came 2011. 'The Arab Spring unnerved Chinese leaders more than any event in years,' commented China observer Evan Osnos.[188] When Egyptian President Hosni Mubarak fell in Cairo in February, the artist and dissident Ai Weiwei tweeted, 'Today, we are all Egyptians.'[189] Protests began to bubble across the country. The response of the Chinese Communist Party was to crack down hard. In April, Ai Weiwei was arrested and detained for 81 days; he was one of 200 people questioned or placed under house arrest.[190] Some were known dissidents, but others included social media celebrities, lawyers, and journalists; anyone with a voice who hinted at threatening the established order.

By the time Xi Jinping became General Secretary of the Chinese Communist Party in 2012, and then President the following year, it was clear that political reform was no longer on the agenda. Instead, the Subject Story would engulf the Consumer, as both technology and consumption became the apparatus of 'stability' – or rather, social control.

By the end of 2020, the government's Skynet project was due to have over 400 million surveillance cameras in place across China, with a growing number of these automatically hooked into facial recognition and other artificial intelligence programmes.[191] The data generated links to the Social Credit System, an enormous data-gathering and -processing exercise which was first announced in 2014, has already been rolled out to millions, and will soon cover the entire nation.[192] Under this system, the Communist Party will know almost everything its citizens are doing, from purchases to driving behaviour to social media posts to the amount of time a person spends playing video games – and will then automatically apply rewards or punishments. One already widespread punishment is to be banned from purchasing flights; according to the National Public Credit Information Centre, this had already happened 17.5 million times by the end of 2018.[193] Other punishments include automatically reduced internet speeds, and even having your pet confiscated. Government documents articulate the underlying principle: 'Keeping trust is glorious and breaking trust is disgraceful.'

It's not hard to see that what 'trust' means here is really obedience, the duty of the Subject. Nor is it hard to see the consequences. Chinese surveillance operations have already been deployed extensively in the state-sponsored persecution of the Uyghur Muslim population of the country and pro-democracy protestors in Hong Kong.

Meanwhile, fuelled by state propaganda, many Chinese people seem broadly happy with the classic Subject bargain of protection, much like those who defended fascism for making the trains run on time. 'I feel like in the past six months, people's behaviour has gotten better and better,' one man said of the impact of the Social Credit System when interviewed by *Foreign Policy* magazine in 2018. 'For example, when we drive, now we always stop in front of crosswalks. If you don't stop, you will lose your points. At first, we just worried about losing points, but now we got used to it.'[194]

The Sovereign Individuals don't just see this as an overseas threat. They believe it represents the alternative to their approach at home in the United States too. If Xi's vision were exported, the Sovereign Individuals could forget about their divine standing; they would end up mere subjects, like the rest of us. Balaji Srinavasaran has explicitly equated the Chinese approach with US attempts to regulate the big tech companies:

In China the recipe was (a) a few years of media demonization plus (b) mandatory Xi Jinping thought sessions followed by (c) decapitation and quasi-nationalization – as is happening with Alibaba and ByteDance. In America it's very similar: (a) a few years of media demonization plus (b) quasi-mandatory wokeness within followed by (c) anti-trust, regulation, and quasi-nationalization... once taken over, these companies will be turned into total surveillance machines and tools of social control. In China perhaps this is already obvious. But in America anti-trust will mean zero trust; in the aftermath of any ostensibly 'economic' settlement the national security state will get everything it ever wanted in terms of backdoors to Google and Facebook. The NSA [National Security Agency] won't need to hack its way in, it'll get a front door.[195]

For Srinavasaran, then, Mark Zuckerberg and his like are fighting a righteous battle against the long arm of the state. In their eyes, 'wokeness' – code for the movements for equality, anti-racism and civil rights – is wrongheaded. They compare it to communism ('if communism is the redistribution of wealth, wokeness is the redistribution of status'[196]), believing it doomed, like communism, to make life far worse for those it claims to champion. The Sovereign Individuals see the proposed break-up of the tech giants as just a disguised pathway to propping up abusive state power, via the spread of the surveillance state.

The irony, though, is that the future Srinavasaran and his like are building would in practice be almost identical to that envisioned by President Xi. Both would represent the Subject Story fusing with the Consumer, if from different starting points. Both would continue the oppression of majority by a minority that both of those stories have enabled. Both would have us believe that the other is the only alternative path forward into the future, and as a result that theirs is the future we must choose.[197]

Yet despite the bandwidth and airwaves devoted to both of

these dystopias, the choice is not down to Sovereign Individual vs China. There is the Citizen Future. The work of transforming the new wave of technology through a Citizen lens is unfolding just as much as it is in all the other domains I have touched on in this book. Campaigners like Tristan Harris of the Center for Humane Technology[198] are winning battles that are starting to reshape the tech giants in the Citizen spirit. Visionary reformers like Mariana Mazzucato, with her concept of *The Entrepreneurial State*,[199] are opening the minds of the establishment to the role governments can play as enablers and establishers of 'missions' for innovation – and how they can work with Citizens to define them (with climate almost always the top priority when they do).

Indeed, the biggest technology news of 2021 has a firmly Citizen feel. The decision of artificial intelligence company DeepMind to make the code of its AlphaFold program entirely open source[200] is arguably the biggest development in science and technology since Tim Berners-Lee did the same with the World Wide Web, and is in many ways analogous: AlphaFold, like the web, represents a fundamental scientific advance, performing predictions of protein structure which could open up all sorts of new opportunities for drug development and much more besides. Rather than control access to this intellectual property, DeepMind has chosen to offer it up as a set of tools that will enable a whole range of advances in public health, just as Berners-Lee did for so many aspects of society. As Azeem Azhar puts it, what is needed in the technology sphere is 'a shift in mindset - one that acknowledges that we have agency over where technology will take us…. Of course, technology builds on what has come before – new innovations layer and combine with those of earlier generations. But its path is not set. We are the ones who decide what we want from the tools we build.'[201] The creator of *Exponential View*, Britain's leading platform for in-depth technology analysis, Azhar is here calling for the conscious adoption of the Citizen Story.

And so the contest for the next dominant story is on. Returning to Ece Temelkuran's metaphor of the reef over the wreck: in every domain, in every institution, each of us faces a deep and personal challenge, not just a prescription or prediction. This challenge applies both to those in positions of power in the existing system, and to the activists working for change from without. The former must not hold on too tightly; they must accept that what they have is a wreck. The city councils of Birmingham and beyond must open their doors and processes to the Immys; the political establishments of Germany and every other nation must allow the Biancas to thrive; the United Nations and the OECD and the European Union must come in behind the Kennedys. If they do not, they will not maintain the status quo: they will simply push us into the arms of the Sovereign Individual libertarians, or the big state authoritarians, or both. Activists and changemakers, for their part, must not reject the wreck entirely, but see it as the starting point for their work, something to which they must respond 'Yes, and...' not 'No, but...' If they do not, if instead they channel anger and rejection, they will be as guilty of herding us into a dystopian future as those in power. This is a challenging realisation for all concerned, one that brings all this right down to a very personal level.

There is no space for heroes in this work.

THE ANTI-HERO

I am white, male, six feet tall, able-bodied, and heterosexual. By the time this book is published, I will be 40. Part of me most definitely wants to be a hero, a role my society has told me I should aspire to. But in the Citizen Story, the era of the exceptional individual, the hero in shining armour, is over.

Heroism has had a good run. By the time the 20th Century wound down, the great majority of us – at least in the West – were living in a hero culture. We were inundated with representations of the individual heroic quest in fiction and film as well as in the news and in history. Not just our leaders and notable achievers but each and every one of us were heroes in our own right. We believed ourselves individually destined to embark on the adventure that would yield the transformation we needed. Yet humans didn't always think of ourselves as heroes. In origin, it's yet another expression of the Subject Story; and heroism's evolution towards universality is yet another expression of the Consumer Story.

The first major scholar of heroes was the Scottish philosopher Thomas Carlyle, who published his thinking *On Heroes, Hero-Worship and the Heroic in History* in 1841.[202] In a perfect reflection of the Subject Story, Carlyle attributed the shaping of the world to Great Men who were born to the task by virtue of their superior intelligence, courage, charisma, and connection to the divine. In his taxonomy, heroes were gods (like Odin or Zeus), kings (like Genghis Khan or Napoleon), prophets or priests (like Muhammed or Martin Luther), and poets or thinkers (like Michelangelo or Descartes). These Great Men made history for us. They deserved our deference; we were merely the inconsequential unnamed extras in their quests.

In all the early myths and epics of Western history, the role of the hero was reserved primarily for immortals. They were superhumans, not relatable figures, with their whole life consisting of extreme acts of conquest and monster-slaying. Upon hearing their stories, common people had no delusions of heroism, of engaging upon a quest of their own. In the medieval period came the chivalric epics featuring brave knights. Although a step closer in relatability than a demigod like Hercules, theirs were still not lives to which everyday people could or did aspire. Beginning in the 1600s, with the tales of explorers (exploiters?) during the Age of Discovery, the role of hero expanded. Tradesmen and the entrepreneurially inclined could embark on colonial adventures. This evolved into the picaresque stories of the 1600s, which portrayed rogues, deviants and outlaws as heroic adventurers, with Don Quixote as the iconic example. Finally, in the 1800s, the notion of adventure itself was broadened, coming to be equated with everyday and/or internal (psychological) struggles rather than just voyages to faraway places or encounters with exotic peoples. Dostoevsky and the American authors Henry James and Edith Wharton, writing in the late 1800s and early 1900s, are considered trailblazers in the genre of psychological fiction.

It was no coincidence that at the same moment as the foundations of the Consumer Story were being laid by the likes of Jevons and Bernays, heroism – previously the terrain of Great Men only – became available to all of us. The rugged individualism and self-reliance of the Consumer Story mapped perfectly onto heroism, as real life converged with the heroic adventure. Over the course of the 20th Century, more and more of us came to think of ourselves as heroes – even, at long last, those of us cast in the roles of victims or enemies for centuries. It became commonplace to project the structure of the hero's journey on our lives, using it to make sense of our choices in hindsight and even predictively to chart our path and set our expectations. We began competing with each other to have the cooler quest, the more exciting adventure, while businesses competed to outfit us with the accessories that would render us victorious. The more privileges we had, the grander those adventures could be. Our social media streams became the platforms for broadcasting our heroic journeys.

Just as with the allure of the Consumer Story's promise of empowerment, the appeal of everyone being the hero of their own adventure is patently obvious. But like the Consumer Story, the cult of heroism is cracking. This, too, is part of the wreck.

On the individual level, the cult of heroism means we are plagued by anxiety, disappointment, and malaise when our outcomes don't match the storybook endings we intended (or worse, broadcasted), like Getting the Girl or the Guy, or Going from Rags to Riches, etc. Heroic culture casts every problem as personal – and the blame for not solving it, a matter of personal responsibility. On the societal level, heroism has us believing in grand gestures and magic solutions, which eclipse the collaborative, interdisciplinary and often messy realities of solving complex problems. This in turn has spawned the saviour complex (which I am vulnerable to), in which people with power believe it's their duty and destiny to rescue 'victims,' those less powerful or less fortunate; with the inevitable consequence, regardless of intent, of further disempowering the victim and instead shoring up the power of the saviour. We have also foisted the role of the hero on people who have no choice in the matter – think of essential workers or healthcare professionals during the pandemic – which might seem like support but is in fact exploitation, an insidious weaponising of heroism.

Finally, even now, despite the democratisation of heroism, despite the diversity training and the vibrant energy around Black Lives Matter or #Time'sUp, the fact remains that what most of us think of as heroic – as leadership, as success, as glory – is still circumscribed by its origins in the Great (White, Straight) Man. We still look to the individual rather than all of us; we admire confidence rather than humility; we expect speaking rather than listening; we welcome rational explanations and data rather than emotions or intuition. Any campaign for a political candidate, any job application for an executive position, any pitch for a new enterprise still highlights these qualities. They are the qualities of a person who has been brought up to think all his ideas are valuable. They're the qualities of the lucky person who gets rewarded by the system at every turn, because the system was designed to reward him, regardless of the quality of his efforts. The person for whom the world-as-it-is functions smoothly, which is to say, a world that holds doors open for whiteness and maleness.

As just one indication of the extent to which people who look and sound like me still dominate, consider this: among the 25,000 or so candidates who stood in the British local elections in 2019, there were more white men named John or David than there were women in total.[203]

This is not to say there aren't countless examples of bold, loud and reasoned people (and heroes, leaders, and successes) who are not well-educated straight white men, or that there are not well-educated straight white men who are shy or plagued by the imposter syndrome, who are empathetic and emotional; but as a rule, the truth of the world is summed up in the meme: 'Lord, grant me the confidence of a mediocre white man.'[204]

The great danger is that everyone who looks like me will react to this truth with defensiveness or guilt. Those shameful and fearful reactions of ours are part of why we're stuck as a society, and they need to be transcended now, in honour of the future. The more time I spend explaining I am one of the good ones, I'm not to blame, I always meant well, I befriend women and people of colour, I too struggled and faced challenges – the less bandwidth and crucial time remains for getting down to work. It is not my fault – or yours, if you resemble me – that I was dealt the high hand of cards I was dealt; however, how I play that hand is my choice. Rather than engage in a macho hunt for moonshots and unicorns, I can choose to play my hand like an anti-hero.

Just as, on a societal level, we must move beyond the Consumer Story, taking with us our independence and uniqueness and reintegrating it into community and collaboration, not disappearing into the powerlessness of the Subject, so there is still value for us in the heroic adventure: the call to leave behind the comforts of the familiar world and enter the unknown; to engage with questions and exploration; to accept the wisdom of mentors; to grapple with doubts and challenges; and to return and integrate the insights and the gifts from the journey, perhaps not always for our own personal benefit, instead for collective benefit. What is different is that the qualities we need in our heroes now are the very opposite of the Great Man's: we need flexibility rather than conviction; we need listening more than speaking; we need emotions and intuition and all the long-underrated sources of intelligence to take their place alongside reason, the rational. We need the 'soft skills' of caring and connecting that get attributed to women's work. We need the sensitivity, flexibility, and adaptation that marginalised people inherently develop because it's how any member of an outgroup survives. These are the qualities of the anti-hero.

The anti-hero must understand everyone else's reality, not just his own. The anti-hero builds coalitions, lifting up points of connection and de-emphasising points of contention, to get things done. The anti-hero can explore multiple possibilities, paths and outcomes, rather than the one story scripted by the status quo. The anti-hero can hold contradictions. The anti-hero is comfortable with the uncertainty inherent in building something radically new and different. Above all, the anti-hero cultivates the instinct to step back and open the door to others, not to step forward and save the day.

In the Citizen Story, the invitation for those of us who have spent our whole lives being listened to and followed is now to listen and to support. Where we have access, we are called to pass along the microphone, to open the doors, and to hand over the keys. We have a vital role to play; it is just not the one we were brought up to anticipate. In many ways, my favourite character of all those I have encountered in the writing of this book is

Speaker Wang: the governing party politician who endorsed and protected the Taiwanese students and hackers, who stood up for them as true democrats. Without him, Taiwan's incredible story would never have been possible. He is my personal role model in all of this, and especially if you look and walk and talk like me, I offer him to you as well.

A FINAL QUESTION

In the course of a meditation on *Hope in the Dark*, Rebecca Solnit writes 'Authentic hope requires clarity – seeing the troubles in this world – and imagination, seeing what might lie beyond these situations that are perhaps not inevitable and immutable.'[205] This clear-eyed yet deeply creative kind of hope is what I have sought in the decade and more since I first framed for myself that traumatic question: What are we doing to ourselves when we tell ourselves we are Consumers 3,000 times a day? I found this hope for myself in a deeper, more rigorous understanding of history and of human nature than I was offered through the best education money can buy. Articulating my second question with Reen and the New Citizenship Project team – What would it look like to put the same energy into inviting participation from people as Citizens? – I then complemented my new-found theory with experiment and practice. With every passing day since, I have become more and more convinced that my hope is justified, even as the problems of the world have intensified.

The kind of hope I have attempted to offer you in this book is rooted in work that can be done from anywhere, in any organisation: opening the doors, inviting people in, creating new ways because we know the old will fail. It is not a hope that puts faith in a panacea or denies the pain and trauma of our current reality, but one that depends on the understanding that this Citizen instinct is already beginning to cohere, that every additional contribution and expression takes us closer to that tipping point – that irresistible critical density of interconnection and relationship – that means that when the Consumer Story fractures again, and the Citizen rises, it cannot and will not be suppressed.

It is a hope that I sustain in my own life by asking myself each and every day the third and final big question I have developed on this journey, which I now offer you as a parting gift:

What would you do in this time, if you truly believed in yourself and those around you?

WRITING CITIZENS: A PARTICIPATORY PROJECT

Like any major undertaking, there have been moments when writing this book has made me want to tear my hair out. But for the most part, I've loved it, not least because it has been an opportunity to practise what I preach, on my own work. In the spirit of the Three Ps, I haven't crowdsourced the content of the book directly; this is after all something I've been working towards for many years. But what I have done is involved people in the process, drawing on ideas, energy and resources from anyone and everyone who was willing to share. It's been a real thrill to work with my own tools in this way – it's made the process much more enjoyable, and I'm also absolutely certain it's made for a better book.

I have had three main sets of collaborators, and in place of a traditional acknowledgements section, I'd like to use this opportunity both to thank them, and to let one or two of them say something about the work that went into this book in their own words. The first is the New Citizenship Project team, and everyone who has been part of it since we started back in 2014. In my mind at least, every project has been both an end in itself, and part of the research for this book. Along the way, every member of the team has taken ownership of the ideas and added their own colour and flavour, and I hope all feel I have done them justice. Reen plays a starring role in the book, and Iris Schönherr, who was there with us at the beginning before moving on to Participatory City, also makes an appearance; but Jo Hunter, Scott Burkholder, Oliver Holtaway, Polly Keane, Katie Dunstan, Emma Ashru Jones, and Tendai Chetse have all shaped the work deeply. I am particularly indebted to Anna Maria Hosford, who has both always supported and often challenged me since we first met working at the National Trust back in 2012; Andy Galloway, whose willingness to get involved and help in any way he could has been matched only by his ability to teach himself to do virtually anything, from building websites (including those for both the book and NCP) to running mass participation events; and Helen Meech, who has been manager, colleague, and client at different times, but always friend above all. But it had to be Reen I asked to offer a few words of her own:

Jon and I reconnected back in 2013, a few years after we'd both left Fallon, a big-name London advertising agency. Jon had been blogging; at the very start of exploring what it would mean to involve people as citizens rather than just sell to them as consumers. At the same time, I had experienced the incredible potential of people power through my own campaigning work on contact lens safety. When we came back together it felt like the perfect meeting of minds: he had been theorising what I had been practising, and each could enrich and build on the other's learning.

When we founded the New Citizenship Project back in 2014, our ambition was to pull the theory and practice together. We aimed to work with the types of organisations we had helped in our advertising days, but this time to help them make what we came to call 'the Citizen Shift' by experimenting with involving people as citizens, standing with the organisation to help deliver its purpose. We deliberately named our company after the title of Jon's blog; we are the New Citizenship Project because this is an ongoing inquiry rather than a fixed set of principles to be applied. Each time we work with a client organisation we learn something new about the Citizen Shift, and Jon has drawn on some of those learnings in researching and writing this book. Over these last seven years, and with a remarkable team alongside us, I am so proud of all we have achieved. And there is so much more to do!

The second of my three main collaborators is my writing partner Ariane Conrad. If the book itself is any good as literature, that is most certainly down to her, and if I have been able to embody the anti-hero I aspire to in the writing, so is that. Most of the best crafted passages and turns of phrase are hers; for those that are mine, she helped me find my voice and stick to it, abandoning the temptation to be 'thesis Jon.' Ariane has worked on many books, but her name has appeared on the covers of very few. If you want an inside track on the big ideas that are going to shape the future of our world, you could do far worse than follow her activities very closely. I'm honoured she was willing to work with me, as she did intensively from early 2020 through to completion in autumn 2021. Over to her:

Had it not been for OuiShare Fest, the annual gathering in Paris of the hearts and minds most inspired by and involved in the early potential of the Collaborative Economy, I would never have met one Jon Alexander, who bounded up to me at the 2016 edition, full of enthusiasm and heart, very much the 'Puppy' his colleagues have aptly dubbed him. Thanks OuiShare!

From there to my book development retreat, the Storia Summit, where I encouraged Jon to expand from a book about 'the Adman's awakening' – the story of the mere genesis of his Citizen formulation, now in the Opening – to this much more comprehensive vision for a whole new Citizen world you hold in your hands.

Not just Jon's expansive, voracious intellect (honed at the finest educational establishments that Whiteness has on offer) but more so his unbounding optimism and deep humility have made this book an utter joy to develop. Our weekly meetings spanned most of the Covid pandemic's first year, and were often my beacon in the turmoil. I do feel as if some of the metaphysical essence of this momentous turning point in human history, which will surely still be decades in the digesting, has pervaded these pages. (Note: Jon graciously tolerated my more magical inclinations. A book, even a heady non-fiction one like ours, is far from all facts, and its creation also involves aspects that I can only describe as mystical. Thanks Universe.)

One thing the pandemic reaffirmed was that books still matter: the business of books has had some banner years as many people, at least those of sufficient means and time to read, searched for consolation and explanation. I believe this one provides both of those things, and I am so proud and honoured to have had a hand in it.

The third main set of collaborators has been my mailing list, invited via my website at www.jonalexander.net. This is where I've been most directly able to put the tools of Part III into practice, sharing the purpose and intention of the book, and inviting people into the process to help me make it real. I've used open source software tools like those that underpin Taiwan's open policymaking processes to generate ideas for the title and subtitle, gathered input to inform and select the cover design, and sent out chapters for review. More importantly though, I've held 'Think Ins' to explore some of the knottiest questions that have come up in the process, openly wrestling with how (and whether) to tell the BrewDog story, and indeed how to promote a book that is essentially anti-consumption! Along the way, I've made new friends myself and brokered new friendships among the list. I've even rediscovered old friends in Tamasine McCaig, Didi Hopkins, and Cathy Runciman who have come forward to help me start plotting the promotion – a stage of the work that is getting under way as this goes to print. I can't overstate how exciting it has been to see the ideas I'm working with play out in reality in such a powerful way. I am proud of this book, and my sincerest hope is that those who have participated in it via that mailing list are, too. One of those who has leaned in most enthusiastically is Ruth Farenga, founder of a consultancy business called Conscious Leaders, and host of an eponymous podcast. Over to her:

I've loved being part of Jon's journey with this book. Jon practises what he preaches and has really embodied a participatory approach in this process. Firstly, this is honourable but moreover, has also borne fruit such as fascinating discussions on the ethics of brands and technology as part of the 'Think In' live events. I really feel that Jon has listened to us and drawn on our input as co-creators.

Jon and I first met when I interviewed him for the Conscious Leaders Podcast. It was then that I realised quite how dedicated he is to the involvement of people in solving the world's most difficult problems; and to the insight that it's when this involvement happens that people move from passive consumers to active stakeholders in solutions. That message has stuck with me and I think Jon does an incredible job, not only at flying the flag for this approach, but also making it real through his work.

Jon's approach will revolutionise the way organisations think about their challenges. For me, it's less 'us and them' and more about collaboration. We don't have time to fight or say that it's someone else's problem and I think Jon helps us bring urgency and proactivity to the challenges of our time.

The thank yous I would like to offer extend way beyond these. I can't name them all, but I consider pretty much everyone I've worked with over the years a collaborator and contributor. Those who have particularly influenced my thinking and work include Craig Mawdsley, Dan Burgess and Jonathan Wise from my time in advertising; Helen Browning (whose farm features in Chapter 5), Jonathan Trimble and Tim Millar (who first came up with the triangle) of those I worked with at the National Trust; Anna Cura and Dan Crossley at the Food Ethics Council; Rich Wilson (the original articulator of the anti-hero) and Claudia Chwalisz from among a whole swathe of fellow travellers in the growing participatory democracy movement; and of course Orit Gal, whose ideas pop up irregularly in this book much as Orit does in my life, and are hopefully as helpful as she always is when they do. A word too for my publisher and editor, Martin Hickman at Canbury Press, who has come in with both serious professional skill and a mind open to my process.

All these have contributed hugely, but without the love and support of my parents and my partner Jane, not only would this book not have been written, but I might not still be here myself. They were there even in those darkest days nearly two decades ago, and have always been there since. My final thank you is to a truly brilliant woman named Chris Seeley, who was my tutor on the Master's in Responsibility and Business Practice at the University of Bath. I started this course while working in advertising, and had left by the time I graduated. Chris held the space for me at a crucial time, as she did for so many, to become who I wanted to be. When she died in late 2014, the world lost one of its wisest, most soulful, and most playful creatures. I will miss her forever.

As a final word, I want to be clear of one more thing: as participatory as the process has been, if there are any errors of any kind here, they are mine alone.

Jon Alexander
November 2021

REFERENCES

1. And that was in 2003. More recent estimates range from 4,000 to 10,000. See for example Jon Simpson, 'Finding Brand Success in the Digital World, *Forbes*, 25 August 2017, https://www.forbes.com/sites/forbesagencycouncil/2017/08/25/finding-brand-success-in-the-digital-world/

2. 'Prime Minister's statement on coronavirus (COVID-19): 10 May 2020,' Prime Minister's Office, https://www.gov.uk/government/speeches/pm-address-to-the-nation-on-coronavirus-10-may-2020

3. 'Coronavirus: Boris Johnson says this is moment of maximum risk,' *BBC News*, 27 April 2020, https://www.bbc.com/news/uk-52439348

4. Haley Ott, '750,000 volunteers answer call to help U.K. health service manage coronavirus crisis,' *CBS News*, 31 March 2020, https://www.cbsnews.com/news/750000-volunteers-answer-call-to-help-u-k-health-service-manage-coronavirus-crisis/

5. Jon Alexander, 'Johnson's message is very deliberate and very dangerous: here's how to combat it,' *Medium*, 10 May, 2020, https://jonjalex.medium.com/johnsons-message-is-very-deliberate-and-very-dangerous-here-s-how-to-combat-it-d336cae96348

6. "Greed' and 'capitalism' helped UK's vaccines success, says PM,' *BBC News*, 24 March 2021, https://www.bbc.com/news/uk-politics-56504546

7. Jenny Kleeman, 'Who Gives?,' *Tortoise*, 16 March 2020, https://www.tortoisemedia.com/2020/03/16/please-give-ungenerously-effective-altruism-who-gives/

The number in MacAskill's book is $3,400. William MacAskill, *Doing Good Better* (New York: Avery, 2015), p. 54.

8. Ibid

9. Ibid

10. p. 20 MacAskill, *Doing Good Better*

11. Rev. Dr Martin Luther King, Jr., *Strength to Love* (Boston: Beacon Press, 1963), p. 25 (in the 1981 edition)

12. Edgar Villanueva, *Decolonizing Wealth* (Oakland: Berrett-Koehler Publishers, 2018), p. 166.

13. Daniela Sirtori-Cortina, 'From Climber To Billionaire: How Yvon Chouinard Built Patagonia Into A Powerhouse His Own Way,' *Forbes*, March 20, 2017 https://www.forbes.com/sites/danielasirtori/2017/03/20/from-climber-to-billionaire-how-yvon-chouinard-built-patagonia-into-a-powerhouse-his-own-way/?sh=15eeb01c275c

14. Lisa Polley, 'Introducing the New Footprint Chronicles' https://www.patagonia.com/stories/introducing-the-new-footprint-chronicles-on-patagoniacom/story-18443.html

15. 'We the Power Campaign Aims to Restore Power to the People,' April 15, 2021, https://www.patagoniaworks.com/press/2021/4/19/we-the-power-campaign-aims-to-restore-power-to-the-people

16. Edward Snowden, 'New Zealand's Prime Minister Isn't Telling the Truth About Mass Surveillance,' *The Intercept*, 15 September 2014, https://theintercept.com/2014/09/15/snowden-new-zealand-surveillance/

17. Bryce Edwards, 'New Zealand's Covid-19 strategy looks successful, but we must safeguard democracy,' the *Guardian*, 15 April 2020, ttps://www.theguardian.com/commentisfree/2020/apr/16/new-zealands-fight-against-covid-19-looks-successful-but-democracy-is-under-threat

18. Eleanor Ainge Roy, "They allowed the perfect storm': UN expert damns New Zealand's housing crisis,' the *Guardian*, 18 Feruary 2020, https://www.theguardian.com/world/2020/feb/19/they-allowed-the-perfect-storm-un-expert-damns-new-zealands-housing-crisis

19. Mark O'Connell, 'Why Silicon Valley billionaires are prepping for the apocalypse in New Zealand,' the *Guardian*, 18 February, 2018, https://www.theguardian.com/news/2018/feb/15/why-silicon-valley-billionaires-are-prepping-for-the-apocalypse-in-new-zealand

20. See http://www.socialacupuncture.co.uk/about/

21. Susan Brenna, 'recovering from Historical Amnesia,' *Teach for America* magazine, 30 January 2018, https://www.teachforamerica.org/stories/recovering-from-historical-amnesia 'That's what locals call a group of highly educated, history-minded millennials who defied a lifetime of warnings from their parents and everyone else, and who returned after college to Stockton to make their stand in the place that raised them.'

22. adrienne maree brown, *Emergent Strategy* (Chico: AK Press, 2017), p. 20.

23. Imandeep Kaur, 'Founding Hub Birmingham,' *Medium*, 20 May 2014, https://medium.com/hub-birmingham/founding-hub-birmingham-ca6db8a251f7

24. 'Mehr als die Hälfte mit ausländischen Wurzeln,' *Frankfurter Allgemeine Zeitung*, 26 June 2017, https://www.faz.net/aktuell/rhein-main/warum-der-auslaenderanteil-in-frankfurt-am-main-so-hoch-ist-15078140.html

25. Rick Falkvinge, *Swarmwise* (Stockholm: CreateSpace Independent Publishing Platform, 2013), pp 295-296

Book available to download at: https://falkvinge.net/files/2013/04/Swarmwise-2013-by-Rick-Falkvinge-v1.1-2013Sep01.pdf

26. Falvinge, p. 14.

27. Alternative fuer Deutschland, 'Manifesto for Germany,' 12 April 2017, https://www.afd.de/wp-content/uploads/sites/111/2017/04/2017-04-12_afd-grundsatzprogramm-englisch_web.pdf

28. Sarah Wildman, 'The German far right is running Islamophobic ads starring women in bikinis,' *Vox*, 31 August 2017, https://www.vox.com/world/2017/8/31/16234008/germany-afd-ad-campaign-far-right

29. Charles Arthur, 'What is the 1% rule?,' the *Guardian*, 20 July 2006, https://www.theguardian.com/technology/2006/jul/20/guardianweeklytechnologysection2

See also: Ben McConnell and Jackie Huba, 'The 1% Rule: Charting citizen participation,' *Church of the Customer* blog, 3 May 2006, https://web.archive.org/web/20100511081141/http://www.churchofthecustomer.com/blog/2006/05/charting_wiki_p.html

30. See https://diem25.org/campaigns/ and https://diem25.org/wp-content/uploads/2016/02/diem25_english_long.pdf

31. Rev. Dr Martin Luther King, Jr., 'The Drum Major Instinct,' Sermon from 4 February, 1968, https://www.youtube.com/watch?v=Mefbog-b4-4

32. Marcus Garvey, speech delivered in Nova Scotia,Canada, 1938, http://marcusgarvey.com/blank-10-2/

33. Kennedy Odede and Jessica Posner, *Find Me Unafraid* (New York: Ecco, 2015), p. 178.

34. Kennedy Odede, 'The pandemic has decimated local economies, where people are more afraid of hunger than the virus,' the *Telegraph*, 13 January 2021, https://www.telegraph.co.uk/global-health/climate-and-people/pandemic-has-decimated-local-economies-people-afraid-hunger/

35. 'Deprivation Index,' UK Local Area, https://www.uklocalarea.com/index.php?l-soa=E01013137&q=East+Marsh&wc=00FCMR

36. Laura Gooderham, 'Grimsby voted the worst place to live in England in 2016,' *Grimsby Telegraph*, 5 January 2017, https://www.grimsbytelegraph.co.uk/news/grims-by-news/grimsby-voted-worst-place-live-70623

37. Mark Townsend, 'How climate change spells disaster for UK fish industry.' the *Guardian*, 26 January 2013, https://www.theguardian.com/environment/2013/jan/26/climate-change-fish-wars-iceland

38. George L. Kelling and James Q. Wilson, 'Broken Windows: The Police and Neighborhood Safety,' *Atlantic*, March 1982.

39. Neither theory went through the rigorous peer-review process that social science requires for theories to gain traction. Nathan J. Robinson, *Superpredator: Bill Clinton's Use and Abuse of Black America* (Philadelphia: Current Affairs Press, 2016). Excerpted in 'Bill Clinton, Superpredator' Jacobin, September 2016. https://www.jacobinmag.com/2016/09/bill-clinton-hillary-superpredators-crime-welfare-african-americans/

40. Eric Klinenberg, 'The Other Side of 'Broken Windows,' *The New Yorker*, 23 August 2018, https://www.newyorker.com/books/page-turner/the-other-side-of-broken-windows

41. David Taylor, 'A Dead Dog in the Bathtub,' *Tortoise*, 11 December 2019, https://www.tortoisemedia.com/2019/12/11/a-dead-dog-in-the-bathtub-thinkin-tour/

42. 'Icelandic Citizen Engagement Tool Offers Tips for U.S.,' The GovLab, 8 January 2020, https://thelivinglib.org/icelandic-citizen-engagement-tool-of-fers-tips-for-u-s/

43. Rob Hopkins, 'A Dazzlingly Delicious Taste of the Future in Liege,' Post Carbon Institute, 28 March 2018, https://www.postcarbon.org/a-dazzlingly-delicious-taste-of-the-future-in-liege/

44. 'Enspiral,' The P2P Foundation Wiki, https://wiki.p2pfoundation.net/Enspiral

45. 'Refugees Designing Future Settlements with Patrick Muvunga,' University of California at Berkeley events, 9 September 2020, https://ced.berkeley.edu/events-media/events/refugees-designing-future-settlements-with-patrick-muvunga

See also: 'Lending to refugees, or giving authors pens: Learnings from the Nakivale Refugee Settlement,' https://www.kiva.org/blog/lending-to-refugees-or-giving-au-thors-pens-learnings-from-the-nakivale-refugee-settlement

46. 'Women Will Defend the Rojava Revolution,' Interview with Zeynep Efrîn, 11 July 2020, https://komun-academy.com/2020/07/11/women-will-defend-the-roja-va-revolution-interview-with-ypj-commander-zeynep-efrin/

See also: Elizabeth Flock, 'Now I've a purpose': why more Kurdish women are choosing to fight,' The *Guardian*, 19 July 2021, https://www.theguardian.com/global-development/20t1/jul/19/came-to-fight-stayed-for-the-freedom-why-more-kurdish-women-are-taking-up-arms

See also: Si Sheppard, 'What the Syrian Kurds Have Wrought,' *The Atlantic*, 25 October 2016, https://www.theatlantic.com/international/archive/2016/10/kurds-rojava-syria-isis-iraq-assad/505037/

47. Ryan Broderick, 'What It's Like To Live-Tweet The Day Your Neighborhood Becomes A War Zone,' *Buzzfeed News*, 30 August 2016, https://www.buzzfeednews.com/article/ryanhatesthis/from-complexo-da-alemao-to-torchbearer

48. Joshua Law, 'How The Coronavirus Is Impacting Favelas In Rio De Janeiro,' *Forbes*, 29 April 2020, https://www.forbes.com/sites/joshualaw/2020/04/29/how-the-coro-navirus-is-impacting-favelas-in-rio-de-janeiro/?sh=6d4fd1d53ee3

See also: 'In Brazil's Favelas, Organizing Is the Difference Between Life and Death,' *Americas Quarterly*, 19 May 2020, https://www.americasquarterly.org/article/in-brazils-favelas-organizing-is-the-difference-between-life-and-death/

49. Dan Nixon, 'Less is more: what does mindfulness mean for economics?,' *Bank Underground*, 25 April 2016, https://bankunderground.co.uk/2016/04/25/less-is-more-what-does-mindfulness-mean-for-economics/

50. Ronald M. Glassman, The Origins of Democracy in Tribes, City-States, and Nation-States (New York: Springer, 2017).

51. Richard Lee, The Dobe Ju/'hoansi (New York: Thomson, 2003), cited in Paul McDowell, 'Political Anthropology: A Cross-Cultural Comparison,' https://socialsci.libretexts.org/Bookshelves/Anthropology/Cultural_Anthropology/Book%3A_Perspectives_-_An_Open_Invitation_to_Cultural_Anthropology/7%3A_Political_Anthropology_-_A_Cross-Cultural_Comparison_(McDowell)/8%3A_Political_Anthropology_-_A_Cross-Cultural_Comparison_(McDowell)

52. Baratunde Thurston, 'Prelude: Revolutionary Love is How to Citizen,' How to Citizen Podcast, 26 August 2020, https://www.baratunde.com/how-to-citizen-episodes/01-revolutionary-love

53. 'Malta golden passports: 'Loopholes' found in citizenship scheme,' *BBC News*, 22 April 2021, https://www.bbc.com/news/world-europe-56843409

54. 'World Migration Report 2020,' International Organization for Migration, 2019, https://www.un.org/sites/un2.un.org/files/wmr_2020.pdf

55. Ibid

56. Ibid

57. Yasmeen Serhan, "Expat' and the Fraught Language of Migration,' *The Atlantic*, 9 October 2018, https://www.theatlantic.com/international/archive/2018/10/expat-immigrant/570967/

58. Stephen Jenkinson, Die Wise (Berkeley: North Atlantic Books, 2015), p. 251.

59. The good news is that even by his strict definition, home and belonging are obviously learnable—as evidenced by Jenkinson's organization Orphan Wisdom, which teaches the skills of deep living and making human culture, 'rooted in knowing history, being claimed by ancestry, working for a time we won't see.' See orphanwisdom.com.

60. Brene Brown, *Braving the Wilderness* (New York: Random House, 2017), p. 37.

61. Marshall McLuhan, *Understanding Media* (New York: McGraw-Hill, 1964).

62. Terry Flew, *New Media: An Introduction* (3rd ed.). (Melbourne: Oxford University Press, 2008), p. 19.

63. Carole Cadwalladr, 'Facebook's Role in Brexit—and the threat to Democracy,' TED2019, April 2019, https://www.ted.com/talks/carole_cadwalladr_facebook_s_role_in_brexit_and_the_threat_to_democracy/transcript

64. Dr Ayana Elizabeth Johnson, 'Is Your Carbon Footprint BS?,' How to Save a Planet podcast, 18 March, 2021, https://gimletmedia.com/shows/howtosaveaplanet/xjh53gn

See also: Dr Ayana Elizabeth Johnson and Dr. Katharine K. Wilkinson, *All We Can Save* (New York: One World, 2020).

65. Dale Dougherty with Ariane Conrad, *Free to Make* (Berkeley: North Atlantic Books, 2016), p. 56

66. Simon Murphy and Clea Skopeliti, 'Coronavirus: campaign launched offering help to those self-isolating,' the *Guardian*, 14 March 2020, https://www.theguardian.com/world/2020/mar/14/coronavirus-campaign-launched-offering-help-to-those-self-isolating

67. Tim Dixon, 'Britain's Choice: us-versus-them, or a bigger 'us'?,' https://www.mhpc.com/britains-choice-us-versus-them-or-a-bigger-us/ based upon research from 'Common Ground and Division in 2020s Britain,' Britain's Choice, https://www.britainschoice.uk/

68. 'NHS volunteer responders: 250,000 target smashed with three quarters of a million committing to volunteer,' *NHS News*, 29 March 2020, https://www.england.nhs.uk/2020/03/250000-nhs-volunteers/

69. Rebecca Solnit, *A Paradise Built in Hell* (New York: Viking, 2009), p. 7.

70. Nichola Raihani, *The Social Instinct* (New York: St. Martin's Press, 2021).

71. Michelle Nijhuis, 'The Miracle of the Commons,' *Aeon*, 4 May 2021, https://aeon.co/essays/the-tragedy-of-the-commons-is-a-false-and-dangerous-myth

72. Rutger Bregman, *Humankind* (New York : Little, Brown and Company, 2020), p. 163

73. Ibid

74. Watch the original ad at https://www.youtube.com/watch?v=2zfqw8nhUwA

75. Marc Bain, 'Fatal muggings for shoes are partially due to sneaker hype, a documentary argues,' *Quartz*, 20 November 2015, https://qz.com/554784/1200-people-are-killed-each-year-over-sneakers/ See also: 'Mom devastated by tragedy pushes for change,' *ABC13*, 21 December 2013, https://abc13.com/archive/9368193/

76. 'The Band Aid Story,' *Channel Four*, https://www.youtube.com/watch?v=BCIMsH_CB_0

77. Eric J. Evans, *Thatcher and Thatcherism* (Abington-on-Thames: Routledge, 1997), p. 27

78. Francis Beckett and David Hencke, *Marching to the Faultline: the Miners' Strike and the Battle for Industrial Britain* (London: Constable, 2009).

79. Ibid.

80. William Stanley Jevons, *Theory of Political Economy* (London: Macmillan and Co., 1871).

81. Edward Bernays, 'The Engineering of Consent,' *The Annals of the American Academy of Political and Social Science*, 1947.

82. 'Schuman declaration May 1950,' European Union/History, https://european-union.europa.eu/principles-countries-history/history-eu/1945-59/schuman-declaration-may-1950_en

83. Economist Victor Lebow writing in The Journal of Retailing in 1955, quoted by Richard Heinberg in his book *Afterburn* (Gabriola Island: New Society Publishers, 2015).

84. George Lakoff, 'Metaphor and War,' *Peace Research*, Issue 23, 1991, pp. 25-32. See https://www.arieverhagen.nl/cms/files/George-Lakoff-1991-Metaphor-and-War.pdf

85. Gitta Sereny, *Albert Speer* (New York: Knopf, 1995).

86. Arlie Russell Hochschild, *Strangers in a Strange Land* (New York: The New Press, 2016).

87. Donella Meadows, 'Leverage Points: Places to Interfere in a System,' originally published 1997, archived at Donella Meadows Institute, https://donellameadows.org/archives/leverage-points-places-to-intervene-in-a-system/

88. James C. Scott, *Against the Grain: A Deep History of the Earliest States* (New Haven: Yale University Press, 2017).

89. 'The Most Influential Figures of Ancient History,' *National Geographic: Special Edition*, 1 January 2016.

90. Jan Morris, *Farewell the Trumpets* (London: Faber & Faber, 1978), pp 21-22.

91. YouGov Survey 2014, https://yougov.co.uk/topics/politics/articles-reports/2014/07/26/britain-proud-its-empire

92. David Cannedine, *The Rise and Fall of the British Aristocracy* (New York: Vintage, 1990), p. 189.

93. Christopher Clark, *The Sleepwalkers* (London: Allen Lane, 2012)

94. Thomas Edward Lawrence, *Seven Pillars of Wisdom: A Triumph* (Private Edition, 1926)

95. Bayo Akomolafe, 'On Slowing Down in Modern Times,' *For the Wild Podcast*, 22 January 2020, https://forthewild.world/podcast-transcripts/https/forthewild-world/listen/bayo-akomolafe-on-slowing-down-in-urgent-times-155

96. Meadows, 'Leverage Points'

97. Ece Temelkuran, *Together: 10 Choices for a Better Now* (London: Fourth Estate Limited, 2021)

98. Thomas Berry 'The New Story,' First published in *Teilhard Studies* no. 1 (winter 1978). Also in *Teilhard in the 21st Century* (Maryknoll, NY: Orbis Books, 2003). http://thomasberry.org/wp-content/uploads/Thomas_Berry-The_New_Story.pdf

99. Thomas Kuhn, *The Structure of Scientific Revolutions* (Chicago: University of Chicago Press, 1962).

100. Akomolafe, 'On Slowing Down in Modern Times'

101. Temelkuran, *Together*

102. 'More than 1 in 3 Americans believe a "deep state" is working to undermine Trump,' NPR/Ipsos poll, 30 December 2020,

https://www.ipsos.com/en-us/news-polls/npr-misinformation-123020

103. Dr Barry Mason, 'Towards Positions of Safe Uncertainty,' *Human Systems: the Journal of Systemic Consultation and Management*, Vol. 4, 1993, pp. 189-200.

104. Ibid.

105. Octavia Hill, *Essays and Letters by Octavia Hill*, edited by Robert Whelan (London: Civitas, 1998), http://www.civitas.org.uk/pdf/rr3.pdf.

106. Additional details about The National Trust from an interview with Dame Fiona Reynolds by author, 5 April, 2021.

107. Katherine Viner, 'A mission for journalism in a time of crisis,' the *Guardian*, 16 November 2017, https://www.theguardian.com/news/2017/nov/16/a-mission-for-journalism-in-a-time-of-crisis

108. Ethan Zuckerman, 'Mistrust, Efficacy and the New Civics,' whitepaper for the Knight Foundation presented at the Aspen Institute, August 2017, https://ethanzuckerman.com/2017/08/17/mistrust-efficacy-and-the-new-civics-a-whitepaper-for-the-knight-foundation/

109. Annie Leonard, 'From Exploitation to Democracy: How We Will Save the Planet,' *The Forge*, 22 July 2020, https://forgeorganizing.org/article/exploitation-democracy-how-we-will-save-planet

110. Ibid

111. Sandra Weiss, Mexico City's 'crowdsourced' constitution,' *International Journal of Politics*, Culture and Society, 27 April 2017, https://www.ips-journal.eu/topics/democracy/mexico-citys-crowdsourced-constitution-1999/

112. Fiona Harvey, 'Global citizens' assembly to be chosen for UN climate talks,' the *Guardian*, 5 October 2021, https://www.theguardian.com/environment/2021/oct/05/global-citizens-assembly-to-be-chosen-for-un-climate-talks

See also: https://globalassembly.org/

113 Will Smale, 'How controversial beer firm BrewDog became so popular,' *BBC News*, 5 January 2015, https://www.bbc.com/news/business-30376484

114. 'Our History,' Brewdog, https://www.brewdog.com/au/community/culture/our-history

115. Paul Mason, 'Brewdog's OpenSource Revolution is at the Vanguard of Postcapitalism,' The *Guardian*, 29 February 2016, https://web.archive.org/web/2016072618t754/https://www.theguardian.com/commentisfree/2016/feb/29/brewdogs-open-source-revolution-is-at-the-vanguard-of-postcapitalism

116. 'Equity Punks and Cicerone,' Brewdog Blog, 16 April 2018, https://www.brewdog.com/blog/equity-punks-and-cicerone

117. James Watt, 'My 10 Best Decisions As BrewDog's CEO,' *LinkedIn*, 30 November 2020, https://www.linkedin.com/pulse/my-10-best-decisions-brewdogs-ceo-james-watt/

118. Ibid

119. 'I AM PUNK,' Brewdog Blog, 25 May 2017, https://blog.brewdog.com/usa/blog/i-am-punk

120. Milton Friedman, *Capitalism and Freedom* (Chicago: University of Chicago Press, 1962).

121. 'How many Certified B Corps are there around the world?,' FAQs, Certified B Corporation, https://bcorporation.net/faq-item/how-many-certified-b-corps-are-there-around-world

122. 'I AM PUNK,' Brewdog Blog

123. See https://buymeonce.com/

124. Author's email correspondence with James Bates-Prince, Head of Brand Strategy, 21 January 2021

125. 'Brewdog [Your Town],' Brewdog Blog, 4 September 2015, https://blog.brewdog.com/usa/blog/brewdog-your-town

126. Ernst-Jan Pfauth, 'Why we see journalists as conversation leaders and readers as expert contributors,' *Medium*, 30 April 2014, https://medium.com/de-correspondent/why-we-see-journalists-as-conversation-leaders-and-readers-as-expert-contributors-8c234ff5bc53

127. Max Falkowitz, 'Birth of the Kale,' *New York Magazine*, 19 April 2018, https://www.grubstreet.com/2018/04/history-of-the-park-slope-food-coop.html

128. See https://www.giffgaff.com/

129. See https://www.yuup.co/

130. Kelly Morris, 'Meet the Entrepreneur Who's Designing Her Company as a Work of Art,' *Huffington Post*, 20 April 2015, https://www.huffpost.com/entry/meet-the-entrepreneur-whos-designing-her-company-as-a-work-of-art_b_7104926

131. 'Brewdog: a Marketing Lesson for Everyone,' Blur Marketing, 3 August 2011, http://blur-marketing.com/blog/brewdog-a-marketing-lesson-for-everyone/

132. Nicola Carruthers, 'BrewDog rectifies trademark row with collaborative gin,' *The Spirits Business*, 21 June 2017, https://www.thespiritsbusiness.com/2017/06/brew-dog-rectifies-trademark-row-with-collaborative-gin/

133. Rob Davies, "Punk' beer maker BrewDog sells 22% of firm to private equity house,' the *Guardian*, 9 April 2017, https://www.theguardian.com/business/2017/apr/09/punk-beermaker-brewdog-sells-22-of-firm-to-private-equity-house

134. Ibid

135. See video at https://twitter.com/BrewDogCamden/status/678949230506045440

136. 'An Open Letter to Brewdog,' 9 June 2021, https://www.punkswithpurpose.org/dearbrewdog/

137. Judith Evans, Alice Hancock and Michael O'Dwyer, 'Punk Rebellion: Brew-Dog's crowdfunding investors start to lose faith,' *Financial Times*, 25 June 2021, https://www.ft.com/content/5ad0e222-a35b-4ae8-aa16-27f1feb964a5

138. 'The Good Future Board,' Good Energy, https://www.goodenergy.co.uk/goodfuture/

139. 'Co-op Group History,' Co-op Legal Services, https://www.co-oplegalservices.co.uk/about-us/the-co-operative-group-history/

140. Josephine Moulds and Jill Treanor, 'Co-op sells farms business to Wellcome Trust,' the *Guardian*, 4 August 2014, https://www.theguardian.com/business/2014/aug/04/co-op-sells-farms-business-wellcome-trust

141. Julia Kollewe and Jill Treanor, 'The Co-operative Bank's timeline of troubles,' The *Guardian*, 11 August 2015, https://www.theguardian.com/business/2015/jun/23/tooperative-bank-timeline-troubles

142. Harriet Sherwood, 'Former Co-op Bank chair Paul Flowers dismissed from church over drugs,' the *Guardian*, 16 January 2017, https://www.theguardian.com/world/2017/jan/16/former-co-op-bank-chair-paul-flowers-dismissed-from-church-over-drugs

143. 'Facts and Figures,' International Cooperative Alliance, https://www.ica.coop/en/cooperatives/facts-and-figures

144. Trebor Scholz, 'Platform Cooperativism vs. the Sharing Economy,' *Medium*, 5 December 2014, https://medium.com/@trebors/platform-cooperativism-vs-the-sharing-economy-2ea737f1b5ad

145. Chris Hughes, 'It's Time to Break Up Facebook,' *New York Times*, 9 May 2019, https://www.nytimes.com/2019/05/09/opinion/sunday/chris-hughes-facebook-zuckerberg.html

146. Details about Taiwan's evolution from interview by the author with Audrey Tang, 19 October, 2020.

147. Paul Mason, *Why It's Kicking Off Everywhere* (London: Verso, 2011).

148. The China Post Staff, 'Cabinet to appoint minister to steer open gov't initiative,' *The China Post*, 1 February 2017.

149. Trisha de Borchgrave, 'How to Handle Covid,' Tortoise, 3 June 2020, https://www.tortoisemedia.com/2020/06/03/how-to-handle-covid/

150. "3 things for Calgary': how a Canadian mayor brought citizens into government," City Monitor, 26 May 2017, https://citymonitor.ai/government/3-things-calgary-how-canadian-mayor-brought-citizens-government-3055

151. See https://reykjavik.is/en/better-reykjavik-0

152. 'Building Consensus and Compromise on Uber in Taiwan,' Centre for Public Impact, 18 September 2019, https://www.centreforpublicimpact.org/case-study/building-consensus-compromise-uber-taiwan

153. Tessy Britton, 'Inviting, welcoming and including everyone.,' *Medium*, 21 February 2021, https://tessybritton.medium.com/inviting-welcoming-and-including-everyone-1bb20df25924

154. Tessy Britton, 'Universal Basic Everything,' *Medium*, 30 May 2020, https://tessybritton.medium.com/universal-basic-everything-f149afc4cef1

155. Rebecca McKee, 'The Citizens' Assembly behind the Irish abortion referendum,' *Involve*, 30 May 2018, https://www.involve.org.uk/resources/blog/opinion/citizens-assembly-behind-irish-abortion-referendum

156. 'Irish abortion law: Citizens' Assembly recommends unrestricted access to terminations,' *BBC News*, 23 April 2017, https://www.bbc.com/news/world-europe-39687584

157. Fintan O'Toole, 'If only Brexit had been run like Ireland's referendum,' the *Guardian*, 29 May 2018, https://www.theguardian.com/commentisfree/2018/may/29/brexit-ireland-referendum-experiment-trusting-people

158. Ibid

159. 'How a False Sense of Security, and a Little Secret Tea, Broke Down Taiwan's COVID-19 Defenses,' *TIME*, 21 May 2021, https://time.com/6050316/taiwan-covid-19-outbreak-tea/

160. Helen Davidson, 'A victim of its own success: how Taiwan failed to plan for a major Covid outbreak,' the *Guardian*, 7 June 2021, https://www.theguardian.com/world/2021/jun/07/a-victim-of-its-own-success-how-taiwan-failed-to-plan-for-a-major-covid-outbreak

161. All Covid case numbers are from John Hopkins COVID-19 tracker, https://systems.jhu.edu/research/public-health/ncov/

162. Kwasi Kwarteng, Priti Patel, Dominic Raab, Chris Skidmore and Elizabeth Truss, *Britannia Unchained* (London: Palgrave MacMillan, 2012)

163. 'Common Ground and Division in 2020s Britain,' Britain's Choice, https://www.britainschoice.uk/

164. John Harris, 'How Flatpack Democracy beat the old parties in the People's Republic of Frome,' the *Guardian*, 22 May 2015, https://www.theguardian.com/politics/2015/may/22/flatpack-democracy-peoples-republic-of-frome

See also: https://www.flatpackdemocracy.co.uk

165. Will Tanner, James O'Shaughnessy, Fjolla Krasniqi, 'The Policies of Belonging,' *Onward*, January 2021, https://www.ukonward.com/wp-content/uploads/2021/01/The-Policies-of-Belonging-1.pdf

166. See https://www.mayorsforagi.org/

167. 'What might a Universal Basic Income mean for Wales?,' Welsh Parliament website, 17 June 2021, https://research.senedd.wales/research-articles/what-might-a-universal-basic-income-mean-for-wales/

168. 'Citizens' assemblies are increasingly popular,' The Economist, 19 September 2020, https://www.economist.com/international/2020/09/17/citizens-assemblies-are-increasingly-popular

169. See https://www.climateassembly.uk/

170. Sandra Laville, 'David Attenborough to appear at citizens' climate assembly,' the *Guardian*, 24 January 2020, https://www.theguardian.com/environment/2020/jan/24/david-attenborough-citizens-climate-assembly

171. 'Game Changer: Creating Canada's Democratic Action Fund,' Mass LBP 2020, https://static1.squarespace.com/static/55af0533e4b04fd6bca65b-c8/t/5e3013a79826d330d5babc27/1580209167185/DAF2020

172. 'Divided and Connected: State of the North 2019,' IPPR North, 27 November 2019, https://www.ippr.org/research/publications/state-of-the-north-2019

Keys findings here: https://www.placenorthwest.co.uk/news/ippr-uk-power-more-centralised-than-any-other-country/

173. Ibid

174. Simon Duffy, 'Colonialism 3.0,' 7 May 2020, https://simonduffy.info/blog/colonialism-version-3

175. Ibid

176. 'Divided and Connected'

177. Ibid

178. Delian Aspourhov on Twitter: https://twitter.com/zebulgar/status/1334991230531301376

179. Arielle Pardes, 'Miami Tech Week Wasn't Planned. But the Hype Is Infectious,' *Wired*, 28 April 2021, https://www.wired.com/story/miami-tech-week-wasnt-planned-but-the-hype-is-infectious/

180. Ibid

181. Douglas Antin, 'Remote Work & The Tech Enabled Exit: Where To Live & Why?,' The Sovereign Individual Weekly (his blog), https://dougantin.com/remote-work-the-tech-enabled-exit/

182. James Dale Davidson and Lord William Rees-Mogg, *The Sovereign Individual* (New York: Simon & Schuster, 1997), p. 18.

183. *The Sovereign Individual*, p. 17.

184. Father of Jacob, currently a minister in Boris Johnson's government, Leader of the House of Commons, and a prominent campaigner for Britain's exit from the European Union.

185. *The Sovereign Individual*, p. 17.

186. Delian Aspourhov on Twitter: https://twitter.com/zebulgar/status/1334991230531301376

187. 'Shopping, or at least browsing, became a principal hobby. The average Chinese citizen was dedicating almost ten hours a week to shopping, while the average American spent less than four.... A study of advertising found that the average person in Shanghai saw three times as many advertisements in a typical day as a consumer in London.' Evan Osnos, *Age of Ambition: Chasing Fortune, Truth and Faith in the New China* (New York: Farrar, Straus and Giroux, 2014).

188. Osnos, *Age of Ambition*

189. Terry Smith, 'Art of Dissent: Ai Weiwei,' *The Monthly*, June 2011, http://www.themonthly.com.au/ai-weiwei-art-dissent-terry-smith-3356

190. Damian Grammaticas, 'Details emerge of Chinese artist Ai Weiwei's detention,' *BBC News*, 11 August 2011, https://www.bbc.com/news/world-asia-pacific-14487328

191. 'In Your Face: China's all-seeing state,' *BBC News*, 10 December 2017, https://www.bbc.com/news/av/world-asia-china-42248056

192. Nicole Kobie, 'The complicated truth about China's social credit system,' *Wired*, 7 June 2019, https://www.wired.co.uk/article/china-social-credit-system-explained

193. Lily Kuo, 'China bans 23m from buying travel tickets as part of 'social credit' system,' the *Guardian*, 1 March 2019, https://www.theguardian.com/world/2019/mar/01/china-bans-23m-discredited-citizens-from-buying-travel-tickets-social-credit-system

194. Simina Mistrau, 'Life Inside China's Social Credit Laboratory,' *Foreign Policy*, 3 April 2018, https://foreignpolicy.com/2018/04/03/life-inside-chinas-social-credit-laboratory/

195. Sotonye, 'If Einstein Had The Internet: An Interview With Balaji Srinivasan,' Time Well Spent substack, 2 August 2021, https://sotonye.substack.com/p/if-einstein-had-the-internet-an-interview

196. Ibid

197. As regards a viable future in terms of climate, to their credit, at least the Chinese acknowledge the climate crisis and have made significant investments in renewable energy and other sustainable technologies. The Sovereign Individuals are significantly more interested in the pipedream of reaching other planets than looking after this one.

198. See https://www.humanetech.com/

199. Mariana Mazzucato, *The Entrepreneurial State* (London: Anthem Press, 2013)

200. Ewen Callaway, 'DeepMind's AI predicts structures for a vast trove of proteins,' *Nature*, 22 July 2021, https://www.nature.com/articles/d41586-021-02025-4

201. Azeem Azhar, *The Exponential Age: How Accelerating Technology Is Leaving Us Behind and What to Do About It* (New York: Diversion Books, 2021).

202. Thomas Carlyle, *On Heroes, Hero- Worship and the Heroic in History* (London: James Fraser, 1841).

203. Daniel Wainwright, 'Council elections: 'Not enough' women and minorities stand,' *BBC News*, 26 April 2019, https://www.bbc.com/news/uk-england-47947867

204. Attributed to Canadian Muslim writer Sarah Hagi.

205. Rebecca Solnit, *Hope in the Dark* (Chicago: Haymarket Books, 2004).

Index

Jon Alexander with Ariane Conrad

314

Scholz, Trebor 230
Schönherr, Iris 253
schools 11
Schuman, Robert 138
scientific methodology 24
Scotland 178, 207, 265, 269, 280
Scott, Ridley 122
secondgov 106
Seedrs 210
selfish gene theory 114
self-reliance 282
Serb nationalism 160
Serendipity engines 48
Sereny, Gitta 145
Seven Modes of Everyday Participation 216, 247
sex workers 97
shanty towns 11
Share connections 218
shareholders 211-212
shareholder value 212
sharing economy 107, 230
Sheffield 267
Sheila McKechnie Award 77
Shining Hope for Communities 65
Shirky, Clay 60
SHOFCO 65, 66, 69-70, 188
Shrewsbury School 141
Sierra Nevada Pale Ale 207
Sikhism 43
Siliconia 6
Silicon Valley 254, 271-272
Silva, Rene 87
Singapore 242
Singer, Peter 24
Sinovac 259
Skynet 276
Sleepwalkers, The 160
Smithsonian 192
Snowden, Edward 32
Snow Revolution 237
social acupuncture 34
Social Credit System 276, 277
social entrepreneurship 87
social media platforms 99
social movements 11
Social order 151
soft skills 108
software program 263
Solnit, Rebecca 113, 286
Solon 91
Somerset 11
sortition 91
South By South West 272
Sovereign Individuals 275-281
Spacehive 247

SpaceX 7, 273
Spanish business group 229
Speer, Albert 144
Sri Lanka 53
Srinavasaran, Balaji 274, 277-278
Stanford Prison Experiment 115
Starbase, Texas 273
startup founders 272
start-up scene 54
startup visas 272
State of the Union Address 202
Stay Alert. Control the Virus. Save Lives. 20
Stay Home. Protect the NHS. Save Lives. 20
Stella Artois 223
Sting 129
Stonehaven 207, 208
Stories 115
Strangers In Their Own Land 145
Structure of Scientific Revolutions, The 167
Suarez, Francis 272
Subject Era 121
Subject Government 246
Subject organisation 180
Subject Story 18, 20-21, 35-36, 148-149, 153-154, 155, 157, 159, 161-163, 171, 173, 179, 201-203, 235, 274, 276-277, 281
Sun and Moon Community Arts Festival 83
Sunday Times, The 208
Sunflower Revolution 238
Super Bowl 122
superbrands 12, 126-127
superpower status, US 138
superpredator theory 81
superwealthy 33
surveillance 275-277
Sustainable Development Goals 250
Swadeshi 159
Swarmwise: the Tactical Manual to Changing the World 55
Sweden 54, 154, 246
Syria 236
Syrian refugees 55

T

tagging 106
Taipei 235, 245, 251, 258, 260

Taiwan 11, 235, 238-239, 241, 243, 245, 248-252, 258, 260-261, 287
Tang, Audrey 239-240, 242, 248, 251
Tate 192
taxation 267-268
tax havens 272
Team Parkinson's 193
tech entrepreneurs 54
Technology Entertainment Design 45
Tech Will Save Us 221
TED Talks 104
TEDxBrum 47
TEDx conference 45
T E Lawrence, Colonel 162
Tell stories 217
Temelkuran, Ece 166-169, 280
Tesco 208
Tesla 273
Thatcher, Margaret 130-131, 133-134, 273-274
Thiel, Peter 33, 274-275
thinkers 281
Third Culture Kids 103
Three Citizen Ps 175
Three Principles of Participatory Organisations 175
Three Ps, The 176
Thurston, Baratunde 95
Tilak, Bal Gangadhar 158
Time Magazine 258
Tipping Point, The 81
Together: 10 Choices for a Better Now 166
Tokyo 267
Tony Blair 12
Tony Blair Faith Foundation 45
Tony's Chocolonely 129
Toronto 266
Tortoise 219, 240
Towards Positions of Safe Uncertainty 170
trade unionism 132
Trespass, Kinder Scout 183, 184
True belonging 104
Trump, Donald 33, 55, 107, 146, 163, 189, 263
Tsai, Jaclyn 239
Tsarist Russia 161
TSG Consumer Partners 223
Tunisia 236
Turkey 53, 166
Turkish family 73
Twitter 223

Canbury

See the world more clearly. All year round.

We're committed to finding talented writers and getting their books into the hands of the right people: people like you. If you'd like to be updated on our new projects, please sign up to our email newsletter at www.canburypress.com – you'll get 10 per cent off all titles you order from our website – or consider our new subscription service.

By subscribing, you allow us to invest in important new non-fiction. For £70 for 12 months (or £5.40 per month), you'll receive:

- Five new books, tailored to you
- With a total recommended retail price of at least £80
- Delivered pre-publication with free P&P
- At least one signed or special edition, or a free gift
- Advance notice and priority booking for author events.

Don't worry – we'll never send you a book you've already ordered from us. To join or gift a subscription, visit: www.canburypress.com/subscriptions

Due to the high cost of postage, this offer is available to UK residents only.